1988

CAMBRIDGE TRACTS IN MATHEMATICS

General Editors

H. BASS, H. HALBERSTAM, J.F.C. KINGMAN
J.E. ROSEBLADE & C.T.C. WALL

85   *The geometry of fractal sets*

*This book is dedicated in affectionate memory of
my Mother and Father*

K. J. FALCONER

Lecturer in Mathematics, University of Bristol
Sometime Fellow of Corpus Christi College, Cambridge

The geometry of fractal sets

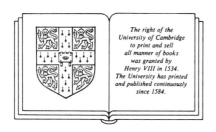

The right of the
University of Cambridge
to print and sell
all manner of books
was granted by
Henry VIII in 1534.
The University has printed
and published continuously
since 1584.

CAMBRIDGE UNIVERSITY PRESS

Cambridge

New York New Rochelle Melbourne Sydney

Published by the Press Syndicate of the University of Cambridge
The Pitt Building, Trumpington Street, Cambridge CB2 1RP
32 East 57th Street, New York, NY 10022, USA
10 Stamford Road, Oakleigh, Melbourne 3166, Australia

First published 1985

First paperback edition (with corrections) 1986
Reprinted 1987
Printed in Great Britain at the University Press, Cambridge

Library of Congress catalogue card number: 84–12091

British Library cataloguing in publication data

Falconer, K.J.
The geometry of fractal sets. – (Cambridge
tracts in mathematics; 85)

1. Geometry   2. Measure theory
I. Title
515.7′3   QA447

ISBN 0 521 25694 1 hard covers
ISBN 0 521 33705 4 paperback

TM

Contents

Preface

This tract provides a rigorous self-contained account of the mathematics of sets of fractional and integral Hausdorff dimension. It is primarily concerned with geometric theory rather than with applications. Much of the contents could hitherto be found only in original mathematical papers, many of which are highly technical and confusing and use archaic notation. In writing this book I hope to make this material more readily accessible and also to provide a useful and precise account for those using fractal sets.

Whilst the book is written primarily for the pure mathematician, I hope that it will be of use to several kinds of more or less sophisticated and demanding reader. At the most basic level, the book may be used as a reference by those meeting fractals in other mathematical or scientific disciplines. The main theorems and corollaries, read in conjunction with the basic definitions, give precise statements of properties that have been rigorously established.

To get a broad overview of the subject, or perhaps for a first reading, it would be possible to follow the basic commentary together with the statements of the results but to omit the detailed proofs. The non-specialist mathematician might also omit the details of Section 1.1 which establishes the properties of general measures from a technical viewpoint.

A full appreciation of the details requires a working knowledge of elementary mathematical analysis and general topology. There is no doubt that some of the proofs central to the development are hard and quite lengthy, but it is well worth mastering them in order to obtain a full insight into the beauty and ingenuity of the mathematics involved.

There is an emphasis on the basic tools of the subject such as the Vitali covering theorem, net measures, and potential theoretic methods.

The properties of measures and Hausdorff measures that we require are established in the first two sections of Chapter 1. Throughout the book the emphasis is on the use of measures in their own right for estimating the size of sets, rather than as a step in defining the integral. Integration is used only as a convenient tool in the later chapters; in the main an intuitive idea of integration should be found perfectly adequate.

Inevitably a compromise has been made on the level of generality adopted. We work in n-dimensional Euclidean space, though many of the ideas apply equally to more general metric spaces. In some cases, where the proofs of higher-dimensional analogues are much more complicated, theorems are only proved in two dimensions, and references are supplied for the extensions. Similarly, one- or two-dimensional proofs are sometimes given if they contain the essential ideas of the general case, but permit simplifications in notation to be made. We also restrict attention to Hausdorff measures corresponding to a numerical dimension s, rather than to an arbitrary function.

A number of the proofs have been somewhat simplified from their original form. Further shortenings would undoubtedly be possible, but the author's desire for perfection has had to be offset by the requirement to finish the book in a finite time!

Although the tract is essentially self-contained, variations and extensions of the work are described briefly, and full references are provided. Further variations and generalizations may be found in the exercises, which are included at the end of each chapter.

It is a great pleasure to record my gratitude to all those who have helped with this tract in any way. I am particularly indebted to Prof Roy Davies for his careful criticism of the manuscript and for allowing me access to unpublished material, and to Dr Hallard Croft for his detailed suggestions and for help with reading the proofs. I am also most grateful to Prof B.B. Mandelbrot, Prof J.M. Marstrand, Prof P. Mattila and Prof C.A. Rogers for useful comments and discussions.

I should like to thank Mrs Maureen Woodward and Mrs Rhoda Rees for typing the manuscript, and also David Tranah and Sheila Shepherd of Cambridge University Press for seeing the book through its various stages of publication. Finally, I must thank my wife, Isobel, for finding time to read an early draft of the book, as well as for her continuous encouragement and support.

Introduction

The geometric measure theory of sets of integral and fractional dimension has been developed by pure mathematicians from early in this century. Recently there has been a meteoric increase in the importance of fractal sets in the sciences. Mandelbrot (1975, 1977, 1982) pioneered their use to model a wide variety of scientific phenomena from the molecular to the astronomical, for example: the Brownian motion of particles, turbulence in fluids, the growth of plants, geographical coastlines and surfaces, the distribution of galaxies in the universe, and even fluctuations of price on the stock exchange. Sets of fractional dimension also occur in diverse branches of pure mathematics such as the theory of numbers and non-linear differential equations. Many further examples are described in the scientific, philosophical and pictorial essays of Mandelbrot. Thus what originated as a concept in pure mathematics has found many applications in the sciences. These in turn are a fruitful source of further problems for the mathematician. This tract is concerned primarily with the geometric theory of such sets rather than with applications.

The word 'fractal' was derived from the latin *fractus*, meaning broken, by Mandelbrot (1975), who gave a 'tentative definition' of a fractal as a set with its Hausdorff dimension strictly greater than its topological dimension, but he pointed out that the definition is unsatisfactory as it excludes certain highly irregular sets which clearly ought to be thought of in the spirit of fractals. Hitherto mathematicians had referred to such sets in a variety of ways – 'sets of fractional dimension', 'sets of Hausdorff measure', 'sets with a fine structure' or 'irregular sets'. Any rigorous study of these sets must also contain an examination of those sets with equal topological and Hausdorff dimension, if only so that they may be excluded from further discussion. I therefore make no apology for including such regular sets (smooth curves and surfaces, etc.) in this account.

Many ways of estimating the 'size' or 'dimension' of 'thin' or 'highly irregular' sets have been proposed to generalize the idea that points, curves and surfaces have dimensions of 0, 1 and 2 respectively. Hausdorff dimension, defined in terms of Hausdorff measure, has the overriding advantage from the mathematician's point of view that Hausdorff measure *is* a measure (i.e. is additive on countable collections of disjoint sets).

Unfortunately the Hausdorff measure and dimension of even relatively simple sets can be hard to calculate; in particular it is often awkward to obtain lower bounds for these quantities. This has been found to be a considerable drawback in physical applications and has resulted in a number of variations on the definition of Hausdorff dimension being adopted, in some cases inadvertently.

Some of these alternative definitions are surveyed and compared with Hausdorff dimension by Hurewicz & Wallman (1941), Kahane (1976), Mandelbrot (1982, Section 39), and Tricot (1981, 1982). They include entropy, see Hawkes (1974), similarity dimension, see Mandelbrot (1982), and the local dimension and measure of Johnson & Rogers (1982). It would be possible to write a book of this nature based on any such definition, but Hausdorff measure and dimension is, undoubtedly, the most widely investigated and the most widely used.

The idea of defining an outer measure to extend the notion of the length of an interval to more complicated sets of real numbers is surprisingly recent. Borel (1895) used measures to estimate the size of sets to enable him to construct certain pathological functions. These ideas were taken up by Lebesgue (1904) as the underlying concept in the construction of his integral. Carathéodory (1914) introduced the more general 'Carathéodory outer measures'. In particular he defined '1-dimensional' or 'linear' measure in n-dimensional Euclidean space, indicating that s-dimensional measure might be defined similarly for other integers s. Hausdorff (1919) pointed out that Carathéodory's definition was also of value for *non-integral s*. He illustrated this by showing that the famous 'middle-third' set of Cantor had positive, but finite, s-dimensional measure if $s = \log 2/\log 3 = 0.6309. \ldots$ Thus the concept of sets of fractional dimension was born, and Hausdorff's name was adopted for the associated dimension and measure.

Since then a tremendous amount has been discovered about Hausdorff measures and the geometry of Hausdorff-measurable sets. An excellent account of the intrinsic measure theory is given in the book by Rogers (1970), and a very general approach to measure geometry may be found in Federer's (1969) scholarly volume, which diverges from us to cover questions of surface area and homological integration theory.

Much of the work on Hausdorff measures and their geometry is due to Besicovitch, whose name will be encountered repeatedly throughout this book. Indeed, for many years, virtually all published work on Hausdorff measures bore his name, much of it involving highly ingenious arguments. More recently his students have made many further major contributions. The obituary notices by Burkill (1971) and Taylor (1975) provide some idea of the scale of Besicovitch's influence on the subject.

It is clear that Besicovitch intended to write a book on geometric measure theory entitled *The Geometry of Sets of Points*, which might well have resembled this volume in many respects. After Besicovitch's death in 1970, Prof Roy Davies, with the assistance of Dr Helen Alderson (who died in 1972), prepared a version of what might have been Besicovitch's 'Chapter 1'. This chapter was not destined to have any sequel, but it has had a considerable influence on the early parts of the present book.

In our first chapter we define Hausdorff measure and investigate its basic properties. We show how to calculate the Hausdorff dimension and measure of sets in certain straightforward cases.

We are particularly interested in sets of dimension s which are s-sets, that is, sets of non-zero but finite s-dimensional Hausdorff measure. The geometry of a class of set restricted only by such a weak condition must inevitably consist of a study of the neighbourhood of a general point. Thus the next three chapters discuss local properties: the density of sets at a point, and the directional distribution of a set round each of its points, that is, the question of the existence of tangents. Sets of fractional and integral dimension are treated separately. Sets of fractional dimension are necessarily fractals, but there is a marked contrast between the regular 'curve-like' or 'surface-like' sets and the irregular 'fractal' sets of integral dimension.

Chapter 5 introduces the powerful technique of net measures. This enables us to show that any set of infinite s-dimensional Hausdorff measure contains an s-set, allowing the theory of s-sets to be extended to more general sets as required. Net measures are also used to investigate the Hausdorff measures of Cartesian products of sets.

The next chapter deals with the projection of sets onto lower-dimensional subspaces. Potential-theoretic methods are introduced as an alternative to a direct geometric approach for parts of this work.

Chapter 7 discusses the 'Kakeya problem, of finding sets of smallest measure inside which it is possible to rotate a segment of unit length. A number of variants are discussed, and the subject is related by duality to the projection theorems of the previous chapter, as well as to harmonic analysis.

The final chapter contains a miscellany of examples that illustrate some of the ideas met earlier in the book.

References are listed at the end of the book and are cited by date. Further substantial bibliographies may be found in Federer (1947, 1969), Rogers (1970) and Mandelbrot (1982).

Notation

With the range of topics covered, particularly in the final chapter, it is impossible to be entirely consistent with the use of notation. In general, symbols are defined when they are first introduced; these notes are intended only as a rough guide.

We work entirely in n-dimensional Euclidean space, $\mathbb{R}^n$. Points of $\mathbb{R}^n$, which are sometimes thought of in the vectorial sense, are denoted by small letters, x, y, z etc. Occasionally we write (x, y) for Cartesian coordinates. Capitals, E, F, Γ, etc. are used for subsets of $\mathbb{R}^n$, and script capitals, $\mathscr{C}, \mathscr{V}, \mathscr{I}$, for families of sets. We use the convention that the set-inclusion symbol $\subset$ allows the possibility of equality. The diameter of the set E is denoted by $|E|$, though, when the sense is clear, the modulus sign also denotes the length of a vector in the usual way, thus $|x - y|$ is the distance between the points x and y. Constants, $b, c, c_1, \varepsilon, \delta$, and indices, i, j, k, are also denoted by lower case letters which may be subscripted.

The following list may serve as a reminder of the notation in more frequent use.

Sets

$\mathbb{R}^n$	n-dimensional Euclidean space.
$B_r(x)$	closed disc or ball, centre x and radius r.
$S_r(x, \theta, \phi)$	sector of angle ϕ and radius r.
$C_r(x, I)$	double sector.
$R(x, y)$	common region of the circle-pair with centres x and y.
$G_{n,k}$	Grossmann manifold of k-dimensional subspaces of $\mathbb{R}^n$.
$L(a, b), L(E)$	line sets.
$\bar{E}, \operatorname{int} E$	topological closure, respectively interior, of E.
$[E]_\delta$	the δ-parallel body to E.

Mappings

$\operatorname{proj}_\theta, \operatorname{proj}_\Pi$	orthogonal projection onto the line in direction θ, resp. the plane Π.
$\hat{f}, \hat{\mu}$	Fourier transforms of the function f and measure μ.
$f \circ g$	composition of the mappings, g followed by f.

Measures etc.

$\mathcal{H}^s$	s-dimensional Hausdorff measure or outer measure.
$\mathcal{L}^n$	n-dimensional Lebesgue measure.
$\mathcal{M}^s$	s-dimensional comparable net measure.
$\mathcal{H}^s_\delta, \mathcal{M}^s_\delta$	δ-outer measures used in constructing $\mathcal{H}^s$ and $\mathcal{M}^s$.
$\mathcal{L}(\Gamma)$	length of the curve Γ.
dim E	Hausdorff dimension of E.
ϕ_t, C_t, I_t	t-potential, capacity, energy.

Densities

$D^s(E, x)$	density of E at x.
$\underline{D}^s(E, x), \bar{D}^s(E, x),$	lower, upper densities.
$\bar{D}^s_c(E, x)$	upper convex density.
$\underline{D}^s(E, x, \theta, \phi)$	lower angular density, etc.

1
Measure and dimension

1.1 Basic measure theory

This section contains a condensed account of the basic measure theory we require. More complete treatments may be found in Kingman & Taylor (1966) or Rogers (1970).

Let X be any set. (We shall shortly take X to be n-dimensional Euclidean space $\mathbb{R}^n$.) A non-empty collection $\mathscr{S}$ of subsets of X is termed a *sigma-field* (or *σ-field*) if $\mathscr{S}$ is closed under complementation and under countable union (so if $E \in \mathscr{S}$, then $X \backslash E \in \mathscr{S}$ and if $E_1, E_2, \ldots \in \mathscr{S}$, then $\bigcup_{j=1}^{\infty} E_j \in \mathscr{S}$). A little elementary set theory shows that a σ-field is also closed under countable intersection and under set difference and, further, that X and the null set $\varnothing$ are in $\mathscr{S}$.

The *lower* and *upper limits* of a sequence of sets $\{E_j\}$ are defined as

$$\varliminf_{j \to \infty} E_j = \bigcup_{k=1}^{\infty} \bigcap_{j=k}^{\infty} E_j$$

and

$$\varlimsup_{j \to \infty} E_j = \bigcap_{k=1}^{\infty} \bigcup_{j=k}^{\infty} E_j.$$

Thus $\varliminf E_j$ consists of those points lying in all but finitely many E_j, and $\varlimsup E_j$ consists of those points in infinitely many E_j. From the form of these definitions it is clear that if E_j lies in the σ-field $\mathscr{S}$ for each j, then $\varliminf E_j$, $\varlimsup E_j \in \mathscr{S}$. If $\varliminf E_j = \varlimsup E_j$, then we write $\lim E_j$ for the common value; this always happens if $\{E_j\}$ is either an increasing or a decreasing sequence of sets.

Let $\mathscr{C}$ be any collection of subsets of X. Then the σ-field *generated by $\mathscr{C}$*, written $\mathscr{S}(\mathscr{C})$, is the intersection of all σ-fields containing $\mathscr{C}$. A straightforward check shows that $\mathscr{S}(\mathscr{C})$ is itself a σ-field which may be thought of as the 'smallest' σ-field containing $\mathscr{C}$.

A *measure* μ is a function defined on some σ-field $\mathscr{S}$ of subsets of X and taking values in the range $[0, \infty]$ such that

$$\mu(\varnothing) = 0 \tag{1.1}$$

and

$$\mu\left(\bigcup_{j=1}^{\infty} E_j\right) = \sum_{j=1}^{\infty} \mu(E_j) \tag{1.2}$$

for every countable sequence of disjoint sets $\{E_j\}$ in $\mathscr{S}$.

It follows from (1.2) that μ is an increasing set function, that is, if $E \subset E'$ and $E, E' \in \mathscr{S}$, then

$$\mu(E) \leq \mu(E').$$

Theorem 1.1 (continuity of measures)
Let μ be a measure on a σ-field $\mathscr{S}$ of subsets of X.
(a) If $E_1 \subset E_2 \subset \ldots$ is an increasing sequence of sets in $\mathscr{S}$, then

$$\mu(\lim_{j \to \infty} E_j) = \lim_{j \to \infty} \mu(E_j).$$

(b) If $F_1 \supset F_2 \supset \ldots$ is a decreasing sequence of sets in $\mathscr{S}$ and $\mu(F_1) < \infty$, then

$$\mu(\lim_{j \to \infty} F_j) = \lim_{j \to \infty} \mu(F_j).$$

(c) For any sequence of sets $\{F_j\}$ in $\mathscr{S}$,

$$\mu(\varliminf_{j \to \infty} F_j) \leq \varliminf_{j \to \infty} \mu(F_j).$$

Proof. (a) We may express $\bigcup_{j=1}^{\infty} E_j$ as the disjoint union
$E_1 \cup \bigcup_{j=2}^{\infty}(E_j \backslash E_{j-1})$. Thus by (1.2),

$$\mu(\lim_{j \to \infty} E_j) = \mu\left(\bigcup_{j=1}^{\infty} E_j \right)$$

$$= \mu(E_1) + \sum_{2}^{\infty} \mu(E_j \backslash E_{j-1})$$

$$= \lim_{k \to \infty} \left[\mu(E_1) + \sum_{2}^{k} \mu(E_j \backslash E_{j-1}) \right]$$

$$= \lim_{k \to \infty} \mu\left(E_1 \cup \bigcup_{2}^{k}(E_j \backslash E_{j-1}) \right)$$

$$= \lim_{k \to \infty} \mu(E_k).$$

(b) If $E_j = F_1 \backslash F_j$, then $\{E_j\}$ is as in (a). Since $\bigcap_j F_j = F_1 \backslash \bigcup_j E_j$,

$$\mu(\lim_{j \to \infty} F_j) = \mu\left(\bigcap_{j=1}^{\infty} F_j \right)$$

$$= \mu(F_1) - \mu(\bigcup_j E_j)$$

$$= \mu(F_1) - \lim_{j \to \infty} \mu(E_j)$$

$$= \lim_{j \to \infty} (\mu(F_1) - \mu(E_j))$$

$$= \lim_{j \to \infty} \mu(F_j).$$

(c) Now let $E_k = \bigcap_{j=k}^{\infty} F_j$. Then $\{E_k\}$ is an increasing sequence of sets in $\mathscr{S}$, so by (a),

$$\mu(\lim_{j \to \infty} F_j) = \mu\left(\bigcup_{k=1}^{\infty} E_k\right) = \lim_{k \to \infty} \mu(E_k) \leq \lim_{j \to \infty} \mu(F_j). \qquad \square$$

Next we introduce outer measures which are essentially measures with property (1.2) weakened to subadditivity. Formally, an *outer measure* v on a set X is a function defined on *all* subsets of X taking values in $[0, \infty]$ such that

$$v(\varnothing) = 0, \tag{1.3}$$

$$v(A) \leq v(A') \quad \text{if } A \subset A' \tag{1.4}$$

and

$$v\left(\bigcup_{1}^{\infty} A_j\right) \leq \sum_{1}^{\infty} v(A_j) \quad \text{for any subsets } \{A_j\} \text{ of } X. \tag{1.5}$$

Outer measures are useful since there is always a σ-field of subsets on which they behave as measures; for reasonably defined outer measures this σ-field can be quite large.

A subset E of X is called *v-measurable* or *measurable with respect to the outer measure v* if it decomposes every subset of X additively, that is, if

$$v(A) = v(A \cap E) + v(A \backslash E) \tag{1.6}$$

for all 'test sets' $A \subset X$. Note that to show that a set E is v-measurable, it is enough to check that

$$v(A) \geq v(A \cap E) + v(A \backslash E), \tag{1.7}$$

since the opposite inequality is included in (1.5). It is trivial to verify that if $v(E) = 0$, then E is v-measurable.

Theorem 1.2

Let v be an outer measure. The collection $\mathscr{M}$ of v-measurable sets forms a σ-field, and the restriction of v to $\mathscr{M}$ is a measure.

Proof. Clearly, $\varnothing \in \mathscr{M}$, so $\mathscr{M}$ is non-empty. Also, by the symmetry of (1.6), $A \in \mathscr{M}$ if and only if $X \backslash A \in \mathscr{M}$. Hence $\mathscr{M}$ is closed under taking complements.

To prove that $\mathscr{M}$ is closed under countable union, suppose that $E_1, E_2, \ldots \in \mathscr{M}$ and let A be any set. Then applying (1.6) to $E_1, E_2, \ldots$ in turn with appropriate test sets,

$$v(A) = v(A \cap E_1) + v(A \backslash E_1)$$
$$= v(A \cap E_1) + v((A \backslash E_1) \cap E_2) + v(A \backslash E_1 \backslash E_2)$$
$$= \cdots$$
$$= \sum_{j=1}^{k} v\left(\left(A \backslash \bigcup_{i=1}^{j-1} E_i\right) \cap E_j\right) + v\left(A \backslash \bigcup_{j=1}^{k} E_j\right).$$

Hence

$$v(A) \geq \sum_{j=1}^{k} v\left(\left(A \backslash \bigcup_{i=1}^{j-1} E_i\right) \cap E_j\right) + v\left(A \backslash \bigcup_{j=1}^{\infty} E_j\right)$$

for all k and so

$$v(A) \geq \sum_{j=1}^{\infty} v\left(\left(A \backslash \bigcup_{i=1}^{j-1} E_i\right) \cap E_j\right) + v\left(A \backslash \bigcup_{j=1}^{\infty} E_j\right). \qquad (1.8)$$

On the other hand,

$$A \cap \bigcup_{j=1}^{\infty} E_j = \bigcup_{j=1}^{\infty} \left(\left(A \backslash \bigcup_{i=1}^{j-1} E_i\right) \cap E_j\right),$$

so, using (1.5),

$$v(A) \leq v\left(A \cap \bigcup_{j=1}^{\infty} E_j\right) + v\left(A \backslash \bigcup_{j=1}^{\infty} E_j\right)$$
$$\leq \sum_{j=1}^{\infty} v\left(\left(A \backslash \bigcup_{i=1}^{j-1} E_i\right) \cap E_j\right) + v\left(A \backslash \bigcup_{j=1}^{\infty} E_j\right) \leq v(A),$$

by (1.8). It follows that $\bigcup_{j=1}^{\infty} E_j \in \mathscr{M}$, so $\mathscr{M}$ is a σ-field.

Now let $E_1, E_2, \ldots$ be *disjoint* sets of $\mathscr{M}$. Taking $A = \bigcup_{j=1}^{\infty} E_j$ in (1.8),

$$v\left(\bigcup_{j=1}^{\infty} E_j\right) \geq \sum_{j=1}^{\infty} v(E_j)$$

and combining this with (1.5) we see that v is a measure on $\mathscr{M}$.   □

We say that the outer measure v is *regular* if for every set A there is a v-measurable set E containing A with $v(A) = v(E)$.

Lemma 1.3
If v is a regular outer measure and $\{A_j\}$ is any increasing sequence of sets,
$$\lim_{j \to \infty} v(A_j) = v(\lim_{j \to \infty} A_j).$$

Proof. Choose a v-measurable E_j with $E_j \supset A_j$ and $v(E_j) = v(A_j)$ for each j. Then, using (1.4) and Theorem 1.1(c),

$$v(\lim A_j) = v(\underline{\lim} A_j) \leq v(\underline{\lim} E_j) \leq \underline{\lim} v(E_j) = \lim v(A_j).$$

The opposite inequality follows from (1.4).   □

Now let (X, d) be a metric space. (For our purposes X will usually be n-dimensional Euclidean space, $\mathbb{R}^n$, with d the usual distance function.) The sets belonging to the σ-field generated by the closed subsets of X are called the *Borel sets* of the space. The Borel sets include the open sets (as complements of the closed sets), the F_σ-*sets* (that is, countable unions of closed sets), the G_δ-*sets* (countable intersections of open sets), etc.

An outer measure v on X is termed a *metric outer measure* if

$$v(E \cup F) = v(E) + v(F) \tag{1.9}$$

whenever E and F are *positively separated*, that is, whenever

$$d(E, F) = \inf\{d(x, y) : x \in E, y \in F\} > 0.$$

We show that if v is a metric outer measure, then the collection of v-measurable sets includes the Borel sets. The proof is based on the following version of 'Carathéodory's lemma'.

Lemma 1.4

Let v be a metric outer measure on (X, d). Let $\{A_j\}_1^\infty$ be an increasing sequence of subsets of X with $A = \lim\limits_{j \to \infty} A_j$, and suppose that $d(A_j, A \backslash A_{j+1}) > 0$ for each j. Then $v(A) = \lim\limits_{j \to \infty} v(A_j)$.

Proof. It is enough to prove that

$$v(A) \leq \lim_{j \to \infty} v(A_j), \tag{1.10}$$

since the opposite inequality follows from (1.4). Let $B_1 = A_1$ and $B_j = A_j \backslash A_{j-1}$ for $j \geq 2$. If $j + 2 \leq i$, then $B_j \subset A_j$ and $B_i \subset A \backslash A_{i-1} \subset A \backslash A_{j+1}$, so B_i and B_j are positively separated. Thus, applying (1.9) $(m-1)$ times,

$$v\left(\bigcup_{k=1}^m B_{2k-1}\right) = \sum_{k=1}^m v(B_{2k-1}),$$

$$v\left(\bigcup_{k=1}^m B_{2k}\right) = \sum_{k=1}^m v(B_{2k}).$$

We may assume that both these series converge – if not we would have $\lim\limits_{j \to \infty} v(A_j) = \infty$, since $\bigcup_{k=1}^m B_{2k-1}$ and $\bigcup_{k=1}^m B_{2k}$ are both contained in A_{2m}. Hence

$$v(A) = v\left(\bigcup_{j=1}^\infty A_j\right) = v\left(A_j \cup \bigcup_{k=j+1}^\infty B_k\right)$$

$$\leq v(A_j) + \sum_{k=j+1}^\infty v(B_k)$$

$$\leq \lim_{i \to \infty} v(A_i) + \sum_{k=j+1}^{\infty} v(B_k).$$

Since the sum tends to 0 as $j \to \infty$, (1.10) follows. $\square$

Theorem 1.5
If v is a metric outer measure on (X,d), then all Borel subsets of X are v-measurable.

Proof. Since the v-measurable sets form a σ-field, and the Borel sets form the smallest σ-field containing the closed subsets of X, it is enough to show that (1.7) holds when E is closed and A is arbitrary.

Let A_j be the set of points in $A \backslash E$ at a distance at least $1/j$ from E. Then $d(A \cap E, A_j) \geq 1/j$, so

$$v(A \cap E) + v(A_j) = v((A \cap E) \cup A_j) \leq v(A) \tag{1.11}$$

for each j, as v is a metric outer measure. The sequence of sets $\{A_j\}$ is increasing and, since E is closed, $A \backslash E = \bigcup_{j=1}^{\infty} A_j$. Hence, provided that $d(A_j, A \backslash E \backslash A_{j+1}) > 0$ for all j, Lemma 1.4 gives $v(A \backslash E) \leq \lim_{j \to \infty} v(A_j)$ and (1.7) follows from (1.11). But if $x \in A \backslash E \backslash A_{j+1}$ there exists $z \in E$ with $d(x,z) < 1/(j+1)$, so if $y \in A_j$ then $d(x,y) \geq d(y,z) - d(x,z) > 1/j - 1/(j+1) > 0$. Thus $d(A_j, A \backslash E \backslash A_{j+1}) > 0$, as required. $\square$

There is another important class of sets which, unlike the Borel sets, are defined explicitly in terms of unions and intersections of closed sets. If (X,d) is a metric space, the *Souslin sets* are the sets of the form

$$E = \bigcup_{i_1 i_2 \ldots} \bigcap_{k=1}^{\infty} E_{i_1 i_2 \ldots i_k},$$

where $E_{i_1 i_2 \ldots i_k}$ is a closed set for each finite sequence $\{i_1, i_2, \ldots, i_k\}$ of positive integers. Note that, although E is built up from a countable collection of closed sets, the union is over continuum-many infinite sequences of integers. (Each closed set appears in the expression in many places.)

It may be shown that every Borel set is a Souslin set and that, if the underlying metric spaces are complete, then any continuous image of a Souslin set is Souslin. Further, if v is an outer measure on a metric space (X,d), then the Souslin sets are v-measurable provided that the closed sets are v-measurable. It follows from Theorem 1.5 that if v is a metric outer measure on (X,d), then the Souslin sets are v-measurable. We shall only make passing reference to Souslin sets. Measure-theoretic aspects are described in greater detail by Rogers (1970), and the connoisseur might also consult Rogers *et al.* (1980).

1.2 Hausdorff measure

For the remainder of this book we work in Euclidean *n*-space, $\mathbb{R}^n$, although it should be emphasized that much of what is said is valid in a general metric space setting.

If U is a non-empty subset of $\mathbb{R}^n$ we define the *diameter* of U as $|U| = \sup\{|x - y| : x, y \in U\}$. If $E \subset \bigcup_i U_i$ and $0 < |U_i| \le \delta$ for each i, we say that $\{U_i\}$ is a δ-*cover* of E.

Let E be a subset of $\mathbb{R}^n$ and let s be a non-negative number. For $\delta > 0$ define

$$\mathscr{H}^s_\delta(E) = \inf \sum_{i=1}^{\infty} |U_i|^s, \qquad (1.12)$$

where the infimum is over all (countable) δ-covers $\{U_i\}$ of E. A trivial check establishes that $\mathscr{H}^s_\delta$ is an outer measure on $\mathbb{R}^n$.

To get the *Hausdorff s-dimensional outer measure* of E we let $\delta \to 0$. Thus

$$\mathscr{H}^s(E) = \lim_{\delta \to 0} \mathscr{H}^s_\delta(E) = \sup_{\delta > 0} \mathscr{H}^s_\delta(E). \qquad (1.13)$$

The limit exists, but may be infinite, since $\mathscr{H}^s_\delta$ increases as δ decreases. $\mathscr{H}^s$ is easily seen to be an outer measure, but it is also a *metric* outer measure. For if δ is less than the distance between positively separated sets E and F, no set in a δ-cover of $E \cup F$ can intersect both E and F, so that

$$\mathscr{H}^s_\delta(E \cup F) = \mathscr{H}^s_\delta(E) + \mathscr{H}^s_\delta(F),$$

leading to a similar equality for $\mathscr{H}^s$. The restriction of $\mathscr{H}^s$ to the σ-field of $\mathscr{H}^s$-measurable sets, which by Theorem 1.5 includes the Borel sets (and, indeed, the Souslin sets) is called *Hausdorff s-dimensional measure*.

Note that an equivalent definition of Hausdorff measure is obtained if the infimum in (1.12) is taken over δ-covers of E by convex sets rather than by arbitrary sets since any set lies in a convex set of the same diameter. Similarly, it is sometimes convenient to consider δ-covers of open, or alternatively of closed, sets. In each case, although a different value of $\mathscr{H}^s_\delta$ may be obtained for $\delta > 0$, the value of the limit $\mathscr{H}^s$ is the same, see Davies (1956). (If however, the infimum is taken over δ-covers by balls, a different measure is obtained; Besicovitch (1928a, Chapter 3) compares such 'spherical Hausdorff measures' with Hausdorff measures.)

For any E it is clear that $\mathscr{H}^s(E)$ is non-increasing as s increases from 0 to ∞. Furthermore, if $s < t$, then

$$\mathscr{H}^s_\delta(E) \ge \delta^{s-t} \mathscr{H}^t_\delta(E),$$

which implies that if $\mathscr{H}^t(E)$ is positive, then $\mathscr{H}^s(E)$ is infinite. Thus there is a unique value, dim E, called the *Hausdorff dimension* of E, such that

$$\mathscr{H}^s(E) = \infty \text{ if } 0 \le s < \dim E, \mathscr{H}^s(E) = 0 \text{ if } \dim E < s < \infty. \qquad (1.14)$$

If C is a cube of unit side in $\mathbb{R}^n$, then by dividing C into k^n subcubes of side $1/k$ in the obvious way, we see that if $\delta \geq k^{-1} n^{\frac{1}{2}}$ then $\mathcal{H}^n_\delta(C) \leq k^n (k^{-1} n^{\frac{1}{2}})^n$ $\leq n^{\frac{1}{2}n}$, so that $\mathcal{H}^n(C) < \infty$. Thus if $s > n$, then $\mathcal{H}^s(C) = 0$ and $\mathcal{H}^s(\mathbb{R}^n) = 0$, since $\mathbb{R}^n$ is expressible as a countable union of such cubes. It follows that $0 \leq \dim E \leq n$ for any $E \subset \mathbb{R}^n$. It is also clear that if $E \subset E'$ then $\dim E \leq \dim E'$.

An $\mathcal{H}^s$-measurable set $E \subset \mathbb{R}^n$ for which $0 < \mathcal{H}^s(E) < \infty$ is termed an *s-set*; a 1-set is sometimes called a *linearly measurable set*. Clearly, the Hausdorff dimension of an *s*-set equals *s*, but it is important to realize that an *s*-set is something much more specific than a measurable set of Hausdorff dimension *s*. Indeed, Besicovitch (1942) shows that any set can be expressed as a disjoint union of continuum-many sets of the same dimension. Most of this book is devoted to studying the geometric properties of *s*-sets.

The definition of Hausdorff measure may be generalized by replacing $|U_i|^s$ in (1.12) by $h(|U_i|)$, where h is some positive function, increasing and continuous on the right. Many of our results have direct analogues for these more general measures, though sometimes at the expense of algebraic simplicity. The Hausdorff 'dimension' of a set E may then be identified more precisely as a partition of the functions which measure E as zero or infinity (see Rogers (1970)). Some progress is even possible if $|U_i|^s$ is replaced by $h(U_i)$, where h is simply a function of the set U_i (see Davies (1969) and Davies & Samuels (1974)).

We next prove that $\mathcal{H}^s$ is a regular measure, together with the useful consequence that we may approximate to *s*-sets from below by closed subsets. This proof is given by Besicovitch (1938) who also demonstrates (1954) the necessity of the finiteness condition in Theorem 1.6(b).

Theorem 1.6

(a) *If E is any subset of $\mathbb{R}^n$ there is a G_δ-set G containing E with $\mathcal{H}^s(G) = \mathcal{H}^s(E)$. In particular, $\mathcal{H}^s$ is a regular outer measure.*

(b) *Any $\mathcal{H}^s$-measurable set of finite $\mathcal{H}^s$-measure contains an F_σ-set of equal measure, and so contains a closed set differing from it by arbitrarily small measure.*

Proof. (a) If $\mathcal{H}^s(E) = \infty$, then $\mathbb{R}^n$ is an open set of equal measure, so suppose that $\mathcal{H}^s(E) < \infty$. For each $i = 1, 2 \ldots$ choose an *open* $2/i$-cover of E, $\{U_{ij}\}_j$, such that

$$\sum_{j=1}^{\infty} |U_{ij}|^s < \mathcal{H}^s_{1/i}(E) + 1/i.$$

Then $E \subset G$, where $G = \bigcap_{i=1}^{\infty} \bigcup_{j=1}^{\infty} U_{ij}$ is a G_δ-set. Since $\{U_{ij}\}_j$ is a $2/i$-cover of G, $\mathcal{H}^s_{2/i}(G) \leq \mathcal{H}^s_{1/i}(E) + 1/i$, and it follows on letting $i \to \infty$ that $\mathcal{H}^s(E) = \mathcal{H}^s(G)$. Since G_δ-sets are $\mathcal{H}^s$-measurable, $\mathcal{H}^s$ is a regular outer measure.

(b) Let E be $\mathcal{H}^s$-measurable with $\mathcal{H}^s(E) < \infty$. Using (a) we may find open sets $O_1, O_2, \ldots$ containing E, with $\mathcal{H}^s(\bigcap_{i=1}^{\infty} O_i \backslash E) = \mathcal{H}^s(\bigcap_{i=1}^{\infty} O_i) - \mathcal{H}^s(E) = 0$. Any open subset of $\mathbb{R}^n$ is an F_σ-set, so suppose $O_i = \bigcup_{i=1}^{\infty} F_{ij}$ for each i, where $\{F_{ij}\}_j$ is an increasing sequence of closed sets. Then by continuity of $\mathcal{H}^s$,

$$\lim_{j \to \infty} \mathcal{H}^s(E \cap F_{ij}) = \mathcal{H}^s(E \cap O_i) = \mathcal{H}^s(E).$$

Hence, given $\varepsilon > 0$, we may find j_i such that

$$\mathcal{H}^s(E \backslash F_{ij_i}) < 2^{-i}\varepsilon \qquad (i = 1, 2, \ldots).$$

If F is the closed set $\bigcap_{i=1}^{\infty} F_{ij_i}$, then

$$\mathcal{H}^s(F) \geq \mathcal{H}^s(E \cap F) \geq \mathcal{H}^s(E) - \sum_{i=1}^{\infty} \mathcal{H}^s(E \backslash F_{ij_i}) > \mathcal{H}^s(E) - \varepsilon.$$

Since $F \subset \bigcap_{i=1}^{\infty} O_i$, then $\mathcal{H}^s(F \backslash E) \leq \mathcal{H}^s(\bigcap_{i=1}^{\infty} O_i \backslash E) = 0$. By (a) $F \backslash E$ is contained in some G_δ-set G with $\mathcal{H}^s(G) = 0$. Thus $F \backslash G$ is an F_σ-set contained in E with

$$\mathcal{H}^s(F \backslash G) \geq \mathcal{H}^s(F) - \mathcal{H}^s(G) > \mathcal{H}^s(E) - \varepsilon.$$

Taking a countable union of such F_σ-sets over $\varepsilon = \frac{1}{2}, \frac{1}{3}, \frac{1}{4}, \ldots$ gives an F_σ-set contained in E and of equal measure to E. $\qquad \square$

The next lemma states that any attempt to estimate the Hausdorff measure of a set using a cover of sufficiently small sets gives an answer not much smaller than the actual Hausdorff measure.

Lemma 1.7

Let E be $\mathcal{H}^s$-measurable with $\mathcal{H}^s(E) < \infty$, and let ε be positive. Then there exists $\rho > 0$, dependent only on E and ε, such that for any collection of Borel sets $\{U_i\}_{i=1}^{\infty}$ with $0 < |U_i| \leq \rho$ we have

$$\mathcal{H}^s(E \cap \bigcup_i U_i) < \sum_i |U_i|^s + \varepsilon.$$

Proof. From the definition of $\mathcal{H}^s$ as the limit of $\mathcal{H}^s_\delta$ as $\delta \to 0$, we may choose ρ such that

$$\mathcal{H}^s(E) < \sum |W_i|^s + \tfrac{1}{2}\varepsilon \qquad (1.15)$$

for any ρ-cover $\{W_i\}$ of E. Given Borel sets $\{U_i\}$ with $0 < |U_i| \leq \rho$, we may find a ρ-cover $\{V_i\}$ of $E \backslash \bigcup_i U_i$ such that

$$\mathcal{H}^s\left(E \backslash \bigcup_i U_i\right) + \tfrac{1}{2}\varepsilon > \sum |V_i|^s.$$

Since $\{U_i\} \cup \{V_i\}$ is then a ρ-cover of E,

$$\mathcal{H}^s(E) < \sum |U_i|^s + \sum |V_i|^s + \tfrac{1}{2}\varepsilon,$$

by (1.15). Hence

$$\mathcal{H}^s\left(E \cap \bigcup_i U_i\right) = \mathcal{H}^s(E) - \mathcal{H}^s\left(E \backslash \bigcup_i U_i\right)$$
$$< \sum |U_i|^s + \sum |V_i|^s + \tfrac{1}{2}\varepsilon - \sum |V_i|^s + \tfrac{1}{2}\varepsilon$$
$$= \sum |U_i|^s + \varepsilon. \qquad \square$$

Finally in this section, we prove a simple lemma on the measure of sets related by a 'uniformly Lipschitz' mapping

Lemma 1.8

Let $\psi : E \to F$ be a surjective mapping such that

$$|\psi(x) - \psi(y)| \le c|x - y| \qquad (x, y \in E)$$

for a constant c. Then $\mathcal{H}^s(F) \le c^s \mathcal{H}^s(E)$.

Proof. For each i, $|\psi(U_i \cap E)| \le c|U_i|$. Thus if $\{U_i\}$ is a δ-cover of E, then $\{\psi(U_i \cap E)\}$ is a $c\delta$-cover of F. Also $\sum_i |\psi(U_i \cap E)|^s \le c^s \sum_i |U_i|^s$ so that $\mathcal{H}^s_{c\delta}(F) \le c^s \mathcal{H}^s_{\delta}(E)$, and the result follows on letting $\delta \to 0$. $\qquad \square$

1.3 Covering results

The Vitali covering theorem is one of the most useful tools of geometric measure theory. Given a 'sufficiently large' collection of sets that cover some set E, the Vitali theorem selects a *disjoint* subcollection that covers almost all of E.

We include the following lemma at this point because it illustrates the basic principle embodied in the proof of Vitali's result, but in a simplified setting. A collection of sets is termed *semidisjoint* if no member of the collection is contained in any different member.

Lemma 1.9

Let $\mathscr{C}$ be a collection of balls contained in a bounded subset of $\mathbb{R}^n$. Then we may find a finite or countably infinite disjoint subcollection $\{B_i\}$ such that

$$\bigcup_{B \in \mathscr{C}} B \subset \bigcup_i B'_i, \qquad\qquad (1.16)$$

where B'_i is the ball concentric with B_i and of five times the radius. Further, we may take the collection $\{B'_i\}$ to be semidisjoint.

Proof. We select the $\{B_i\}$ inductively. Let $d_0 = \sup\{|B| : B \in \mathscr{C}\}$ and choose B_1 from $\mathscr{C}$ with $|B_1| \ge \tfrac{1}{2}d_0$. If $B_1, \ldots, B_m$ have been chosen let $d_m = \sup\{|B| : B \in \mathscr{C}, B \text{ disjoint from } \bigcup_1^m B_i\}$. If $d_m = 0$ the process terminates. Otherwise choose B_{m+1} from $\mathscr{C}$ disjoint from $\bigcup_1^m B_i$ with $|B_{m+1}| \ge \tfrac{1}{2}d_m$. Certainly, these balls are disjoint; we claim that they also have the required covering property. If $B \in \mathscr{C}$, then either $B = B_i$ for some i, or B intersects some B_i with $2|B_i| \geq |B|$. If this was not the case B would have been selected in preference to the first ball B_m for which $2|B_m| < |B|$. (Note that, by summing volumes, $\sum |B_i|^2 < \infty$ so that $|B_i| \to 0$ as $i \to \infty$ if infinitely many balls are selected.) In either case, $B \subset B_i'$, giving (1.16). To get the $\{B_i'\}$ semidisjoint, simply remove B_i from the subcollection if $B_i' \subset B_j'$ for any $j \neq i$ noting that B_i' can only be contained in finitely many B_j'. $\qquad \square$

A collection of sets $\mathcal{V}$ is called a *Vitali class* for E if for each $x \in E$ and $\delta > 0$ there exists $U \in \mathcal{V}$ with $x \in U$ and $0 < |U| \leq \delta$.

Theorem 1.10 (Vitali covering theorem)
(a) *Let E be an $\mathcal{H}^s$-measurable subset of $\mathbb{R}^n$ and let $\mathcal{V}$ be a Vitali class of closed sets for E. Then we may select a (finite or countable) disjoint sequence $\{U_i\}$ from $\mathcal{V}$ such that either $\sum |U_i|^s = \infty$ or $\mathcal{H}^s(E \backslash \bigcup_i U_i) = 0$.*
(b) *If $\mathcal{H}^s(E) < \infty$, then, given $\varepsilon > 0$, we may also require that*

$$\mathcal{H}^s(E) \leq \sum_i |U_i|^s + \varepsilon.$$

Proof. Fix $\rho > 0$; we may assume that $|U| \leq \rho$ for all $U \in \mathcal{V}$. We choose the $\{U_i\}$ inductively. Let U_1 be any member of $\mathcal{V}$. Suppose that $U_1, \ldots, U_m$ have been chosen, and let d_m be the supremum of $|U|$ taken over those U in $\mathcal{V}$ which do not intersect $U_1, \ldots, U_m$. If $d_m = 0$, then $E \subset \bigcup_1^m U_i$ so that (a) follows and the process terminates. Otherwise let U_{m+1} be a set in $\mathcal{V}$ disjoint from $\bigcup_1^m U_i$ such that $|U_{m+1}| \geq \frac{1}{2} d_m$.

Suppose that the process continues indefinitely and that $\sum |U_i|^s < \infty$. For each i let B_i be a ball with centre in U_i and with radius $3|U_i|$. We claim that for every $k > 1$

$$E \backslash \bigcup_1^k U_i \subset \bigcup_{k+1}^\infty B_i. \qquad (1.17)$$

For if $x \in E \backslash \bigcup_1^k U_i$ there exists $U \in \mathcal{V}$ not intersecting $U_1, \ldots, U_k$ with $x \in U$. Since $|U_i| \to 0$, $|U| > 2|U_m|$ for some m. By virtue of the method of selection of $\{U_i\}$, U must intersect U_i for some i with $k < i < m$ for which $|U| \leq 2|U_i|$. By elementary geometry $U \subset B_i$, so (1.17) follows. Thus if $\delta > 0$,

$$\mathcal{H}_\delta^s \left(E \backslash \bigcup_1^\infty U_i \right) \leq \mathcal{H}_\delta^s \left(E \backslash \bigcup_1^k U_i \right) \leq \sum_{k+1}^\infty |B_i|^s = 6^s \sum_{k+1}^\infty |U_i|^s,$$

provided k is large enough to ensure that $|B_i| \leq \delta$ for $i > k$. Hence $\mathcal{H}_\delta^s(E \backslash \bigcup_1^\infty U_i) = 0$ for all $\delta > 0$, so $\mathcal{H}^s(E \backslash \bigcup_1^\infty U_i) = 0$, which proves (a).

To get (b), we may suppose that ρ chosen at the beginning of the proof is the number corresponding to ε and E given by Lemma 1.7. If $\sum |U_i|^s = \infty$, then (b) is obvious. Otherwise, by (a) and Lemma 1.7,

$$\mathcal{H}^s(E) = \mathcal{H}^s(E \backslash \bigcup_i U_i) + \mathcal{H}^s(E \cap \bigcup_i U_i)$$

$$= 0 + \mathcal{H}^s(E \cap \bigcup_i U_i)$$

$$< \sum |U_i|^s + \varepsilon. \qquad \square$$

Covering theorems are studied extensively in their own right, and are of particular importance in harmonic analysis, as well as in geometric measure theory. Results for very general classes of sets and measures are described in the two books by de Guzmán (1975, 1981) which also contain further references. One approach to covering principles is due to Besicovitch (1945*a*, 1946, 1947); the first of these papers includes applications to densities such as described in Section 2.2 of this book.

1.4 Lebesgue measure

We obtain *n*-dimensional Lebesgue measure as an extension of the usual definition of the volume in $\mathbb{R}^n$ (we take 'volume' to mean length in $\mathbb{R}^1$ and area in $\mathbb{R}^2$).

Let C be a coordinate block in $\mathbb{R}^n$ of the form

$$C = [a_1, b_1) \times [a_2, b_2) \times \cdots \times [a_n, b_n),$$

where $a_i < b_i$ for each i. Define the volume of C as

$$V(C) = (b_1 - a_1)(b_2 - a_2) \ldots (b_n - a_n)$$

in the obvious way. If $E \subset \mathbb{R}^n$ let

$$\mathcal{L}^n(E) = \inf \sum_i V(C_i), \tag{1.18}$$

where the infimum is taken over all coverings of E by a sequence $\{C_i\}$ of blocks. It is easy to see that $\mathcal{L}^n$ is an outer measure on $\mathbb{R}^n$, known as *Lebesgue n-dimensional outer measure*. Further, $\mathcal{L}^n(E)$ coincides with the volume of E if E is any block; this follows by approximating the sum in (1.18) by a finite sum and then by subdividing E by the planes containing the faces of the C_i. Since any block C_i may be decomposed into small subblocks leaving the sum in (1.18) unaltered, it is enough to take the infimum over δ-covers of E for any $\delta > 0$. Thus $\mathcal{L}^n$ is a metric outer measure on $\mathbb{R}^n$. The restriction of $\mathcal{L}^n$ to the $\mathcal{L}^n$-measurable sets or *Lebesgue-measurable sets*, which, by Theorem 1.5, include the Borel sets, is called *Lebesgue n-dimensional measure* or *volume*.

Clearly, the definitions of $\mathcal{L}^1$ and $\mathcal{H}^1$ on $\mathbb{R}^1$ coincide. As might be expected, the outer measures $\mathcal{L}^n$ and $\mathcal{H}^n$ on $\mathbb{R}^n$ are related if $n > 1$, in fact they differ only by a constant multiple. To show this we require the following well-known geometric result, the 'isodiametric inequality', which says that the set of maximal volume of a given diameter is a sphere. Proofs, using symmetrization or other methods, may be found in any text on convexity, e.g. Eggleston (1958), see also Exercise 1.6.

Theorem 1.11

The n-dimensional volume of a closed convex set of diameter d is, at most, $\pi^{\frac{1}{2}n}(\frac{1}{2}d)^n/(\frac{1}{2}n)!$, the volume of a ball of diameter d.

Theorem 1.12

If $E \subset \mathbb{R}^n$, then $\mathscr{L}^n(E) = c_n \mathscr{H}^n(E)$, where $c_n = \pi^{\frac{1}{2}n}/2^n(\frac{1}{2}n)!$. In particular, $c_1 = 1$ and $c_2 = \pi/4$.

Proof. Given $\varepsilon > 0$ we may cover E by a collection of closed convex sets $\{U_i\}$ such that $\sum |U_i|^n < \mathscr{H}^n(E) + \varepsilon$. By Theorem 1.11 $\mathscr{L}^n(U_i) \le c_n |U_i|^n$, so $\mathscr{L}^n(E) \le \sum \mathscr{L}^n(U_i) < c_n \mathscr{H}^n(E) + c_n \varepsilon$, giving $\mathscr{L}^n(E) \le c_n \mathscr{H}^n(E)$.

Conversely, let $\{C_i\}$ be a collection of coordinate blocks covering E with

$$\sum_i V(C_i) < \mathscr{L}^n(E) + \varepsilon. \tag{1.19}$$

We may suppose these blocks to be open by expanding them slightly whilst retaining this inequality. For each i the closed balls contained in C_i of radius, at most, δ form a Vitali class for C_i. By the Vitali covering theorem, Theorem 1.10(a), there exist disjoint balls $\{B_{ij}\}_j$ in C_i of diameter, at most, δ, with $\mathscr{H}^n(C_i \backslash \bigcup_{j=1}^{\infty} B_{ij}) = 0$ and so with $\mathscr{H}_\delta^n(C_i \backslash \bigcup_{j=1}^{\infty} B_{ij}) = 0$. Since $\mathscr{L}^n$ is a Borel measure, $\sum_{j=1}^{\infty} \mathscr{L}^n(B_{ij}) = \mathscr{L}^n(\bigcup_{j=1}^{\infty} B_{ij}) \le \mathscr{L}^n(C_i)$. Thus

$$\mathscr{H}_\delta^n(E) \le \sum_{i=1}^{\infty} \mathscr{H}_\delta^n(C_i) \le \sum_{i=1}^{\infty} \sum_{j=1}^{\infty} \mathscr{H}_\delta^n(B_{ij}) + \sum_{i=1}^{\infty} \mathscr{H}_\delta^n\left(C_i \backslash \bigcup_{j=1}^{\infty} B_{ij} \right)$$

$$\le \sum_{i=1}^{\infty} \sum_{j=1}^{\infty} |B_{ij}|^n = \sum_{i=1}^{\infty} \sum_{j=1}^{\infty} c_n^{-1} \mathscr{L}^n(B_{ij})$$

$$\le c_n^{-1} \sum_{i=1}^{\infty} \mathscr{L}^n(C_i) < c_n^{-1} \mathscr{L}^n(E) + c_n^{-1} \varepsilon,$$

by (1.19). Thus $c_n \mathscr{H}_\delta^n(E) \le \mathscr{L}^n(E) + \varepsilon$ for all ε and δ, giving $c_n \mathscr{H}_\delta^n(E) \le \mathscr{L}^n(E)$. $\square$

One of the classical results in the theory of Lebesgue measure is the Lebesgue density theorem. Much of our later work stems from attempts to formulate such a theorem for Hausdorff measures. The reader may care to furnish a proof as an exercise in the use of the Vitali covering theorem. Alternatively, the theorem is a simple consequence of Theorem 2.2.

Theorem 1.13 (Lebesgue density theorem)

Let E be an $\mathscr{L}^n$-measurable subset of $\mathbb{R}^n$. Then the Lebesgue density of E at x,

$$\lim_{r \to 0} \frac{\mathscr{L}^n(E \cap B_r(x))}{\mathscr{L}^n(B_r(x))}, \tag{1.20}$$

exists and equals 1 if $x \in E$ and 0 if $x \notin E$, except for a set of x of $\mathscr{L}^n$-measure 0. ($B_r(x)$ denotes the closed ball of centre x and radius r, and, as always, r tends to 0 through positive values.)

1.5 Calculation of Hausdorff dimensions and measures

It is often difficult to determine the Hausdorff dimension of a set and harder still to find or even to estimate its Hausdorff measure. In the cases that have been considered it is usually the lower estimates that are awkward to obtain. We conclude this chapter by analysing the dimension and measure of certain sets; further examples will be found throughout the book. It should become apparent that there is a vast range of s-sets in $\mathbb{R}^n$ for all values of s and n, so that the general theory to be described is widely applicable.

The most familiar set of real numbers of non-integral Hausdorff dimension is the Cantor set. Let $E_0 = [0, 1]$, $E_1 = [0, 1/3] \cup [2/3, 1]$, $E_2 = [0, 1/9] \cup [2/9, 1/3] \cup [2/3, 7/9] \cup [8/9, 1]$, etc., where E_{j+1} is obtained by removing the (open) middle third of each interval in E_j; see Figure 1.1. Then E_j consists of 2^j intervals, each of length 3^{-j}. Cantor's set is the perfect (closed and dense in itself) set $E = \bigcap_{j=0}^{\infty} E_j$. (The collection of closed intervals that occur in this construction form a 'net', that is, any two such intervals are either disjoint or else one is contained in the other. The idea of a net of sets crops up frequently in this book.) Equivalently, E is, to within a countable set of points, the set of numbers in [0, 1] whose base three expansions do not contain the digit 1. We calculate explicitly the Hausdorff dimension and measure of E; this basic type of computation extends to rather more complicated sets.

Fig. 1.1

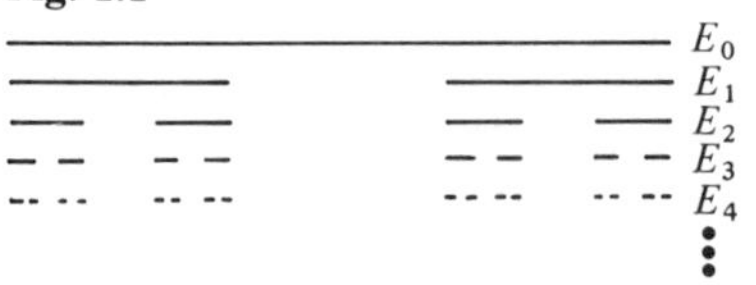

Theorem 1.14

The Hausdorff dimension of the Cantor set E is $s = \log 2/\log 3 = 0.6309\ldots$. Moreover, $\mathscr{H}^s(E) = 1$.

Proof. Since E may be covered by the 2^j intervals of length 3^{-j} that form

E_j, we see at once that $\mathscr{H}^s_{3^{-j}}(E) \le 2^j 3^{-sj} = 2^j 2^{-j} = 1$. Letting $j \to \infty$, $\mathscr{H}^s(E) \le 1$.

To prove the opposite inequality we show that if $\mathscr{I}$ is any collection of intervals covering E, then

$$1 \le \sum_{I \in \mathscr{I}} |I|^s. \tag{1.21}$$

By expanding each interval slightly and using the compactness of E, it is enough to prove (1.21) when $\mathscr{I}$ is a finite collection of closed intervals. By a further reduction we may take each $I \in \mathscr{I}$ to be the smallest interval that contains some pair of net intervals, J and J', that occur in the construction of E. (J and J' need not be intervals of the same E_j.) If J and J' are the largest such intervals, then I is made up of J, followed by an interval K in the complement of E, followed by J'. From the construction of the E_j we see that

$$|J|, |J'| \le |K|. \tag{1.22}$$

Then

$$|I|^s = (|J| + |K| + |J'|)^s$$
$$\ge (\tfrac{3}{2}(|J| + |J'|))^s = 2(\tfrac{1}{2}|J|^s + \tfrac{1}{2}|J'|^s) \ge |J|^s + |J'|^s,$$

using the concavity of the function t^s and the fact that $3^s = 2$. Thus replacing I by the two subintervals J and J' does not increase the sum in (1.21). We proceed in this way until, after a finite number of steps, we reach a covering of E by equal intervals of length 3^{-j}, say. These must include all the intervals of E_j, so as (1.21) holds for this covering it holds for the original covering $\mathscr{I}$. □

There is nothing special about the factor $\tfrac{1}{3}$ used in the construction of the Cantor set. If we let E_0 be the unit interval and obtain E_{j+1} by removing a proportion $1 - 2k$ from the centre of each interval of E_j, then by an argument similar to the above (with (1.22) replaced by $|J|, |J'| \le |K|k/(1 - 2k)$ we may show that $\mathscr{H}^s(\bigcap_1^\infty E_j) = 1$, where $s = \log 2/\log(1/k)$.

We may construct irregular subsets in higher dimensions in a similar fashion. For example, take E_0 to be the unit square in $\mathbb{R}^2$ and delete all but the four corner squares of side k to obtain E_1. Continue in this way, at the jth stage replacing each square of E_{j-1} by four corner squares of side k^j to get E_j (see Figure 1.2 for the first few stages of construction). Then the same sort of calculation gives positive upper and lower bounds for $\mathscr{H}^s(\bigcap_1^\infty E_j)$, where $s = \log 4/\log(1/k)$. More precision is required to find the exact value of the measure in such cases, and we do not discuss this further.

Instead, we describe a generalization of the Cantor construction on the

Fig. 1.2

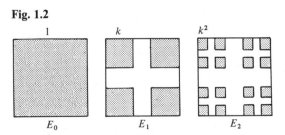

real line. Let s be a number strictly between 0 and 1; the set constructed will have dimension s. Let E_0 denote the unit interval; we define inductively sets $E_0 \supset E_1 \supset E_2...$, each a finite union of closed intervals, by specifying $E_{j+1} \cap I$ for each interval I of E_j. If I is such an interval, let $m \geq 2$ be an integer, and let $J_1, J_2, ..., J_m$ be equal and equally spaced closed subintervals of I with lengths given by

$$|J_i|^s = \frac{1}{m}|I|^s, \tag{1.23}$$

and such that the left end of J_1 coincides with the left end of I and the right end of J_m with the right end of I. Thus

$$m|J_i| + (m-1)d = |I| \quad (1 \leq i \leq m), \tag{1.24}$$

where d is the spacing between two consecutive intervals J_i. Define E_{j+1} by requiring that $E_{j+1} \cap I = \bigcup_1^m J_i$. Note that the value of m may vary over different intervals I in E_j, so that the sets E_j can contain intervals of many different lengths.

The set $E = \bigcap_{j=0}^{\infty} E_j$ is a perfect nowhere dense subset of the unit interval. The following analysis is to appear in a forthcoming paper of Davies.

Theorem 1.15

If E is the set described above, then $\mathcal{H}^s(E) = 1$.

Proof. An interval used in the construction of E, that is, a component subinterval of some E_j, is called a net interval. For $F \subset E$ let

$$\mu(F) = \inf \sum_{I \in \mathcal{I}} |I|^s, \tag{1.25}$$

where the infimum is taken over all possible coverings of F by collections $\mathcal{I}$ of net intervals. Then μ is an outer measure (and, indeed, a Borel measure) on the subsets of E. Note that the value of μ is unaltered if we insist that $\mathcal{I}$ be a δ-cover of F for case $\delta > 0$, since using (1.23) we may always replace a net interval I by a number of smaller net intervals without altering the sum in (1.25).

Let $\mathcal{I}$ be a cover of E by net intervals. To find a lower bound for $\sum_{I \in \mathcal{I}} |I|^s$ we may assume that the collection $\mathcal{I}$ is finite (since each net interval is open relative to the compact set E) and also that the intervals in $\mathscr{I}$ are pairwise disjoint (we may remove those intervals contained in any others by virtue of the net property). Let J be one of the shortest intervals of $\mathscr{I}$; suppose that J is a component interval of E_j, say. Then $J \subset I$ for some interval I in E_{j-1}. Since $\mathscr{I}$ is a disjoint cover of E, all the other intervals of $E_j \cap I$ must be in $\mathscr{I}$. If we replace these intervals by the single interval I, the value of $\sum |I|^s$ is unaltered by (1.23). We may proceed in this way, replacing sets of net intervals by larger intervals without altering the value of the sum, until we reach the single interval $[0, 1]$. It follows that $\sum_{I \in \mathscr{I}} |I|^s = |[0, 1]|^s = 1$, so, in particular,

$$\mu(E) = 1.$$

In exactly the same manner we see that if J is any net interval, then

$$\mu(J \cap E) = |J|^s. \tag{1.26}$$

Next we show that

$$\mu(J \cap E) \leq |J|^s \tag{1.27}$$

for an *arbitrary* interval J. Contracting J if necessary, it is enough to prove this on the assumptions that $J \subset [0, 1]$ and that the endpoints of J lie in E, and, by approximating, coincide with endpoints of net intervals contained in J. Let I be the smallest net interval containing J; say I is an interval of E_j. Suppose that J intersects the intervals $J_q, J_{q+1}, \ldots, J_r$ among the component intervals of $E_{j+1} \cap I$, where $1 \leq q < r \leq m$. (There must be at least two such intervals by the minimality of I.) We claim that

$$|J_q \cap J|^s + |J_{q+1}|^s + \cdots + |J_{r-1}|^s + |J_r \cap J|^s \leq |J|^s. \tag{1.28}$$

If $J_q \cap J$ is not the whole of J_q or if $J_r \cap J$ is not the whole of J_r, then on increasing J slightly the left-hand side of inequality (1.28) increases faster than the right-hand side. Hence it is enough to prove (1.28) when J is the smallest interval containing J_q and J_r. Under such circumstances (1.28) becomes

$$k|J_i|^s \leq |J|^s = (k|J_i| + (k-1)d)^s, \tag{1.29}$$

where $k = r - q + 1$. This is true if $k = m$ by (1.23) and (1.24), and is trivial if $k = 1$, with equality holding in both cases. Differentiating twice, we see that the right-hand expression of (1.29) is a convex function of k, so (1.29) holds for $1 \leq k \leq m$, and the validity of (1.28) follows.

Finally, if either $J_q \cap J$ or $J_r \cap J$ is not a single net interval, we may repeat the process, replacing $J_q \cap J$ and $J_r \cap J$ by smaller net intervals to obtain an expression similar to (1.28) but involving intervals of E_{j+2} rather than of E_{j+1}. We continue in this way to find eventually that $|J|^s$ is at least the sum of the sth powers of the lengths of disjoint net intervals covering $J \cap E$ and contained in J. Thus (1.27) follows from (1.26) for any interval J.

As (1.25) remains true if the infimum is taken over δ-covers $\mathcal{I}$ for any $\delta > 0$, $\mathcal{H}^s(E) \leq \mu(E)$. On the other hand, by (1.27),

$$\mu(E) \leq \sum \mu(J_i \cap E) \leq \sum |J_i|^s$$

for any cover $\{J_i\}$ of E, so $\mu(E) \leq \mathcal{H}^s(E)$. We conclude that $\mathcal{H}^s(E) = \mu(E) = 1$. $\square$

Similar constructions in higher dimensions involve nested sequences of squares or cubes rather than intervals. The same method allows the Hausdorff dimension to be found and the corresponding Hausdorff measure to be estimated.

The basic method of Theorem 1.15 may also be applied to find the dimensions of other sets of related types. For example, if in the construction of E the intervals $J_1, \ldots, J_m$ in each I are just 'nearly equal' or 'nearly equally spaced', the method may be adapted to find the dimension of E. Similarly, if in obtaining E_{j+1} from E_j equations (1.23) and (1.24) only hold 'in the limit as $j \to \infty$', it may still be possible to find the dimension of E.

Another technique useful for finding the dimension of a set is to 'distort' it slightly to give a set of known dimension and to apply Lemma 1.8. The reader may wish to refer to Theorem 8.15(a) where this is illustrated.

Eggleston (1952) finds the Hausdorff dimension of very general sets formed by intersection processes; his results have been generalized by Peyrière (1977). Recently an interesting and powerful method has been described by Davies & Fast (1978). Other related constructions are given by Randolph (1941), Erdös (1946), Ravetz (1954), Besicovitch & Taylor (1954), Beardon (1965), Best (1942), Cigler & Volkmann (1963) and Wegmann (1971*b*), these last three papers continuing earlier works of the same authors. A further method of estimating Hausdorff measures is described in Section 8.3.

Exercises on Chapter 1

1.1   Show that if μ is a measure on a σ-field of sets $\mathcal{M}$ and $E_j \in \mathcal{M}(1 \leq j < \infty)$, then $\mu(\overline{\lim_{j \to \infty}} E_j) \geq \overline{\lim_{j \to \infty}} \mu(E_j)$ provided that $\mu(\bigcup_1^\infty E_j) < \infty$.

1.2   Let v be an outer measure on a metric space (X, d) such that every Borel set is v-measurable. Show that v is a metric outer measure.

1.3   Show that the outer measure $\mathcal{H}^s$ on $\mathbb{R}^n$ is translation invariant, that is, $\mathcal{H}^s(x + E) = \mathcal{H}^s(E)$, where $x + E = \{x + y : y \in E\}$. Deduce that $x + E$ is $\mathcal{H}^s$-measurable if and only if E is $\mathcal{H}^s$-measurable. Similarly, show that $\mathcal{H}^s(cE) = c^s \mathcal{H}^s(E)$, where $cE = \{cy : y \in E\}$.

1.4   Prove the following version of the Vitali covering theorem for a general measure μ: let E be a μ-measurable subset of $\mathbb{R}^n$ with $\mu(E) < \infty$. If $\mathcal{V}$ is a

Vitali class of (measurable) sets for E, then there exist disjoint sets $U_1, U_2, \ldots \in \mathscr{V}$ such that $\mu(E \backslash \bigcup_i U_i) = 0$.

1.5 Use the Vitali covering theorem to prove the Lebesgue density theorem. (Consider the class of balls $\mathscr{V} = \{B_r(x) : x \in E, r \leq \rho \text{ and } \mathscr{L}^n(B_r(x) \cap E) \leq \alpha \mathscr{L}^n(B_r(x))\}$ for each $\alpha < 1$ and $\rho > 0$.)

1.6 Prove that the area of a plane convex set U of diameter d is, at most, $\frac{1}{4}\pi d^2$. (For one method take a point on the boundary of U as origin for polar coordinates so that the area of U is $\frac{1}{2}\int r(\phi)^2 \, d\phi$, and observe that $r(\phi)^2 + r(\phi + \frac{1}{2}\pi)^2 \leq d^2$ for each ϕ.)

1.7 Use the Lebesgue density theorem to deduce the result of Steinhaus, that if E is a Lebesgue-measurable set of real numbers of positive measure, then the difference set $\{y - x : x, y \in E\}$ contains an interval $(-h, h)$. Show more generally that if E and E' are measurable with positive Lebesgue measure, then $\{y - x : x \in E, y \in E'\}$ contains an interval.

1.8 Let μ be a Borel measure on $\mathbb{R}^n$ and let E be a μ-measurable set with $0 < \mu(E) < \infty$. Show that

(a) if $\varlimsup_{r \to 0} r^{-s} \mu(B_r(x) \cap E) < c < \infty$ for $x \in E$, then $\mathscr{H}^s(E) > 0$,

(b) if $\varlimsup_{r \to 0} r^{-s} \mu(B_r(x) \cap E) > c > 0$ for $x \in E$, then $\mathscr{H}^s(E) < \infty$.

(For (a) use the definition of Hausdorff measure, for (b) use the version of the Vitali covering theorem in Exercise 1.4.)

1.9 Let E be the set of numbers between 0 and 1 that contain no odd digit in their decimal expansion. Obtain the best upper and lower estimates that you can for the Hausdorff dimension and measure of E. (In fact E is an s-set where $s = \log 5/\log 10$. This example is intended to illustrate some of the difficulties that can arise in finding Hausdorff measures, being a little more awkward than the Cantor set. One approach to such questions is described in Section 8.3.)

2

Basic density properties

2.1 Introduction

Recall that a subset E of $\mathbb{R}^n$ is termed an s-set ($0 \leq s \leq n$) if E is $\mathscr{H}^s$-measurable and $0 < \mathscr{H}^s(E) < \infty$. In general we exclude the case of $s = 0$ as this often requires separate treatment from other values of s. However, since 0-sets are simply finite sets of points their properties are straightforward.

The next three chapters are concerned with local properties of s-sets, in particular with questions of density and the existence of tangents. As measure properties carry over under countable unions, some of the results may be adapted for measurable sets of σ-finite $\mathscr{H}^s$-measure (i.e. sets formed as countable unions of s-sets). Sets of dimension s of non-σ-finite $\mathscr{H}^s$-measure are difficult to get any sort of hold on except by finding subsets of finite measure, and this is discussed in Chapter 5.

Usually, s will be regarded as fixed and, where there is no ambiguity, terms such as 'measure', 'measurable' and 'almost all' (i.e. 'except for a set of measure zero') refer to the measure $\mathscr{H}^s$.

First we define the basic set densities that play a major rôle in our development. These densities are natural analogues of Lebesgue density (1.20), though their behaviour can be very different. The densities are indicative of the local measure of a set compared with the 'expected' measure.

Let $B_r(x)$ denote the closed ball of centre x and radius r so that $|B_r(x)| = 2r$. The *upper* and *lower densities* (sometimes called upper and lower *spherical* or *circular densities*) of an s-set E at a point $x \in \mathbb{R}^n$ are defined as

$$\bar{D}^s(E, x) = \varlimsup_{r \to 0} \frac{\mathscr{H}^s(E \cap B_r(x))}{(2r)^s}$$

and

$$\underline{D}^s(E, x) = \varliminf_{r \to 0} \frac{\mathscr{H}^s(E \cap B_r(x))}{(2r)^s}$$

respectively. If $\bar{D}^s(E, x) = \underline{D}^s(E, x)$ we say that the density of E at x exists and we write $D^s(E, x)$ for the common value.

A point $x \in E$ at which $\underline{D}^s(E, x) = \bar{D}^s(E, x) = 1$ is called a *regular* point of E; otherwise x is an *irregular* point. An s-set E is said to be *regular* if $\mathscr{H}^s$-almost all of its points are regular and *irregular* if almost all of its points are irregular.

Characterizing regular s-sets and obtaining bounds for the densities of s-sets are two important aims of this book. There is no analogue of the Lebesgue density theorem; it is not in general true that an s-set has density 1 at almost all of its points. In fact one of the main results of the subject is that an s-set cannot be regular unless s is an integer. If s is integral, however, an s-set decomposes into a regular and an irregular part. Very roughly speaking, a regular s-set looks like a measurable subset of an s-dimensional manifold in $\mathbb{R}^n$, whereas an irregular set might behave as a Cartesian product of n Cantor-like sets chosen so that the resulting set is of the required dimension.

Whilst we are mainly concerned with spherical densities, it is also convenient to introduce the *upper convex density* of an s-set E at x, defined as

$$\bar{D}_c^s(E, x) = \lim_{r \to 0} \left\{ \sup \frac{\mathscr{H}^s(E \cap U)}{|U|^s} \right\}, \tag{2.1}$$

where the supremum is over all *convex* sets U with $x \in U$ and $0 < |U| \le r$. Any set is contained in a convex set of equal diameter, so this is equivalent to taking the supremum over *all* sets U with $x \in U$ and $0 < |U| \le r$.

Since $B_r(x)$ is convex and since if $x \in U$, then $U \subset B_r(x)$ where $r = |U|$, we have the relations

$$2^{-s} \bar{D}_c^s(E, x) \le \bar{D}^s(E, x) \le \bar{D}_c^s(E, x). \tag{2.2}$$

Before we prove the basic results on densities, it may be of value to enunciate some general principles that are frequently applied.

First, we often need to know that sets of points defined in terms of metric or measure properties are measurable with respect to an appropriate measure. In practice in this subject, checking measurability is nearly always a formality. The sets encountered can usually be expressed in terms of known measurable sets using combinations of $\underline{\lim}$, $\overline{\lim}$, countable unions and intersections, etc. Or again, we may wish to consider $\bigcap_{\rho > 0} E_\rho$, say, but find on examining the definition of the sets E_ρ that this is the same as the *countable* intersection $\bigcap_{\rho \in \mathbb{Q}^+} E_\rho$ over positive rational values of ρ. Some demonstrations of measurability are given in Lemma 2.1 but, subsequently, when a routine check suffices, we often assume without explicit mention that the sets involved are measurable.

Second, many proofs involve showing that a set of points defined by a property such as a density condition has measure zero. Typically, we need to show that the set $E = \{x : \psi(x) > 0\}$ has Hausdorff measure zero, where ψ is some real-valued function. A standard approach is to show that for each

$\alpha > 0$ the set $\{x : \psi(x) > \alpha\}$ has measure zero and to point out that the same must then be true of the countable union $E = \bigcup_{j=1}^{\infty} \{x : \psi(x) > 1/j\}$. Thus it is enough to prove that $\mathcal{H}^s\{x : \psi(x) > \alpha\} < \varepsilon$ for all $\alpha, \varepsilon > 0$.

Third, we mention the idea of 'uniformization'. If we require a property to hold at almost all points of a set E, it is enough to show that, for all $\varepsilon > 0$, there is a subset F of E with the measure of $E \backslash F$ less than ε and with the property holding throughout F. The advantage of this is that F can often be chosen to behave much more regularly than E. For example, F can generally be taken as a closed set using Theorem 1.6, and might also be chosen so that various densities, for example, converge uniformly on F. Throughout the book, we generally use the letter F to denote this 'working set', obtained by stripping E of its most violent irregularities. Very often if a result can be proved for a set under reasonable topological and uniformity assumptions it is a purely technical matter to extend the result to full generality.

2.2 Elementary density bounds

In this section we commence our study of fundamental density properties of s-sets by proving some results which are valid for all values of s. Such results were first given for linearly measurable sets by Besicovitch (1928a, 1938), the latter paper containing improved proofs. There is no difficulty in generalizing this work to s-sets in $\mathbb{R}^n$ for any s and n, as was indicated by Besicovitch (1945a), Marstrand (1954a) and Wallin (1969).

First we check the measurability of the densities as functions of x. It is sometimes simpler to talk about the measurability of a function rather than of sets defined by that function. The function $f : \mathbb{R}^n \to \mathbb{R}$ is *measurable* (resp. *Borel-measurable, upper semicontinuous*) if the set $\{x : f(x) < c\}$ is a measurable (resp. Borel, open set for every c); an equivalent definition of measurability is obtained if '$<$' is replaced by '$\leq$', '$>$' or '$\geq$'. It follows from the open set definition of continuity that continuous functions on $\mathbb{R}^n$ are Borel measurable.

Lemma 2.1

Let E be an s-set.
(a) *$\mathcal{H}^s(E \cap B_r(x))$ is an upper semicontinuous and so Borel-measurable function of x for each r.*
(b) *$\underline{D}^s(E, x)$ and $\bar{D}^s(E, x)$ are Borel-measurable functions of x.*

Proof. (a) Given $r, \alpha > 0$ write

$$F = \{x : \mathcal{H}^s(E \cap B_r(x)) < \alpha\}.$$

Let $x \in F$. As $\varepsilon \searrow 0$, then $B_{r+\varepsilon}(x)$ decreases to $B_r(x)$, so by the continuity of $\mathcal{H}^s$ from above,

$$\mathcal{H}^s(E \cap B_{r+\varepsilon}(x)) \searrow \mathcal{H}^s(E \cap B_r(x)).$$

Thus we may find ε such that $\mathcal{H}^s(E \cap B_{r+\varepsilon}(x)) < \alpha$, so if $|y - x| \leq \varepsilon$, then $B_r(y) \subset B_{r+\varepsilon}(x)$ and $\mathcal{H}^s(E \cap B_r(y)) < \alpha$. Hence F is an open subset of $\mathbb{R}^n$. This is true for all α, so we conclude that $\mathcal{H}^s(E \cap B_r(x))$ is upper semicontinuous in x.

(b) Using part (a),

$$\{x : \mathcal{H}^s(E \cap B_r(x)) < \alpha(2r)^s\}$$

is open, so, given $\rho > 0$,

$$F_\rho = \{x : \mathcal{H}^s(E \cap B_r(x)) < \alpha(2r)^s \text{ for some } r < \rho\}$$

is the union of such sets and so is open. Now

$$\{x : \underline{D}^s(E, x) < \alpha\} = \bigcap_{\rho > 0} F_\rho;$$

since F_ρ increases as ρ decreases, we may take this intersection over the countable set of positive rational values of ρ. Hence $\{x : \underline{D}^s(E, x) < \alpha\}$ is a G_δ-, and so a Borel set for each α, making $\underline{D}^s(E, x)$ a Borel-measurable function of x. A similar argument establishes the measurability of $\bar{D}^s(E, x)$. $\square$

This lemma enables us to assert that sets such as $\{x : \underline{D}^s(E, x) > \alpha\}$ are $\mathcal{H}^s$-measurable for any s-set E. A further consequence of the proof is that sets of the form

$$\{x : \mathcal{H}^s(E \cap B_r(x)) < \alpha(2r)^s \text{ for some } r \leq \rho\}$$

are $\mathcal{H}^s$-measurable (as open sets); a minor variation allows 'some' to be replaced by 'all'.

We obtain bounds for upper densities first in the case of *convex* densities, and then use (2.2) to deduce the corresponding results for circular densities. The following theorem is obvious if E is a compact set, but is true more generally.

Theorem 2.2
If E is an s-set in $\mathbb{R}^n$, then $\bar{D}^s_c(E, x) = 0$ for $\mathcal{H}^s$-almost all $x \notin E$.

Proof. Fix $\alpha > 0$; we show that the measurable set

$$F = \{x \notin E : \bar{D}^s_c(E, x) > \alpha\}$$

has zero measure. By the regularity of $\mathcal{H}^s$ we may, given $\delta > 0$, find a closed $E_1 \subset E$ with $\mathcal{H}^s(E \backslash E_1) < \delta$. For $\rho > 0$ let

$$\mathcal{V} = \{U : U \text{ is closed and convex}, 0 < |U| \leq \rho, U \cap E_1 = \varnothing$$
$$\text{and } \mathcal{H}^s(E \cap U) > \alpha |U|^s\}. \tag{2.3}$$

Then $\mathcal{V}$ is a Vitali class of closed sets for F, using (2.1) and the closure of E_1, and so we may use the Vitali theorem, Theorem 1.10(a), to find a disjoint sequence of sets $\{U_i\}$ in $\mathcal{V}$ with either $\sum |U_i|^s = \infty$ or $\mathcal{H}^s(F \backslash \bigcup U_i) = 0$.

But by (2.3),

$$\sum |U_i|^s < \frac{1}{\alpha} \sum \mathcal{H}^s(E \cap U_i) = \frac{1}{\alpha} \mathcal{H}^s(E \cap \bigcup_i U_i)$$

$$\leq \frac{1}{\alpha} \mathcal{H}^s(E \backslash E_1) < \frac{\delta}{\alpha} < \infty,$$

as the $\{U_i\}$ are disjoint and are disjoint from E_1. We conclude that $\mathcal{H}^s(F \backslash \cup U_i) = 0$, so that

$$\mathcal{H}_\rho^s(F) \leq \mathcal{H}_\rho^s(F \backslash \bigcup U_i) + \mathcal{H}_\rho^s(F \cap \bigcup U_i)$$

$$\leq \mathcal{H}^s(F \backslash \bigcup U_i) + \sum |U_i|^s < 0 + \frac{\delta}{\alpha}.$$

This is true for any $\delta > 0$ and any $\rho > 0$, so $\mathcal{H}^s(F) = 0$, as required. $\square$

Theorem 2.3
If E is an s-set in $\mathbb{R}^n$, then $\bar{D}_c^s(E, x) = 1$ at $\mathcal{H}^s$-almost all $x \in E$.

Proof. (a) We use the definition of Hausdorff measure to show that $\bar{D}_c^s(E, x) \geq 1$ almost everywhere in E. Take $\alpha < 1$ and $\rho > 0$ and let

$$F = \{x \in E : \mathcal{H}^s(E \cap U) < \alpha |U|^s \text{ for all convex}$$

$$U \text{ with } x \in U \text{ and } |U| \leq \rho\}. \qquad (2.4)$$

Then F is a Borel subset of E. For any $\varepsilon > 0$ we may find a ρ-cover of F by convex sets $\{U_i\}$ such that

$$\sum |U_i|^s < \mathcal{H}^s(F) + \varepsilon.$$

Hence, assuming that each U_i contains some point of F, and using (2.4),

$$\mathcal{H}^s(F) \leq \sum_i \mathcal{H}^s(F \cap U_i) \leq \sum_i \mathcal{H}^s(E \cap U_i)$$

$$< \alpha \sum_i |U_i|^s < \alpha(\mathcal{H}^s(F) + \varepsilon).$$

Since $\alpha < 1$ and the outer inequality holds for all $\varepsilon > 0$, we conclude that $\mathcal{H}^s(F) = 0$. We may define such an F for any $\rho > 0$, so, by definition, $\bar{D}_c^s(E, x) \geq \alpha$ for almost all $x \in E$. This is true for all $\alpha < 1$, so we conclude that $\bar{D}_c^s(E, x) \geq 1$ almost everywhere in E.
(b) We use a Vitali method to show that $\bar{D}_c^s(E, x) \leq 1$ almost everywhere. Given $\alpha > 1$, let $F = \{x \in E : \bar{D}_c^s(E, x) > \alpha\}$, so that F is a measurable subset of E. Let $F_0 = \{x \in F : \bar{D}_c^s(E \backslash F, x) = 0\}$. Then $\mathcal{H}^s(F \backslash F_0) = 0$ by Theorem 2.2. Using the definition of convex density, $\bar{D}_c^s(F, x) \geq \bar{D}_c^s(E, x) - \bar{D}_c^s(E \backslash F, x) > \alpha$ if $x \in F_0$.
Thus

$$\mathcal{V} = \{U : U \text{ is closed and convex and}$$

$$\mathcal{H}^s(F \cap U) > \alpha |U|^s\} \qquad (2.5)$$

is a Vitali class for F_0, so by Theorem 1.10(b) we may, given $\varepsilon > 0$, find a disjoint sequence of sets $\{U_i\}_i$ in $\mathscr{V}$ with $\mathscr{H}^s(F_0) \le \sum |U_i|^s + \varepsilon$. By (2.5)

$$\mathscr{H}^s(F) = \mathscr{H}^s(F_0) < \frac{1}{\alpha} \sum \mathscr{H}^s(F \cap U_i) + \varepsilon \le \frac{1}{\alpha} \mathscr{H}^s(F) + \varepsilon.$$

This inequality holds for any $\varepsilon > 0$, so $\mathscr{H}^s(F) = 0$ if $\alpha > 1$, as required. $\qquad\square$

Results akin to Theorems 2.2 and 2.3 have been obtained by Freilich (1966) and Davies & Samuels (1974) for surprisingly general measures of Hausdorff type.

The analogues of these two theorems for circular densities, which are rather more important in our development, follow immediately using (2.2).

Corollary 2.4
If E is an s-set in $\mathbb{R}^n$, then $D^s(E, x) = 0$ at $\mathscr{H}^s$-almost all x outside E.

Corollary 2.5
If E is an s-set in $\mathbb{R}^n$, then

$$2^{-s} \le \bar{D}^s(E, x) \le 1$$

at almost all $x \in E$.

Corollary 2.4 has a number of important consequences. The first of these is that the densities of a measurable subset of an s-set coincide with the densities of the original set at almost all points of the subset. This result is often used in a preparatory manner; thus when examining a subset with a certain density property we can discard the remainder of the set leaving the density property holding almost everywhere.

Corollary 2.6
Let F be a measurable subset of an s-set E. Then $\underline{D}^s(F, x) = \underline{D}^s(E, x)$ and $\bar{D}^s(F, x) = \bar{D}^s(E, x)$ for almost all $x \in F$.

Proof.
Writing $H = E \backslash F$ we have from Corollary 2.4 that $D^s(H, x) = 0$ at almost all x in F. For such x

$$\underline{D}^s(E, x) = \underline{D}^s(F, x) + D^s(H, x) = \underline{D}^s(F, x) \text{ and}$$
$$\bar{D}^s(E, x) = \bar{D}^s(F, x) + D^s(H, x) = \bar{D}^s(F, x). \qquad\square$$

Corollary 2.7
Let $E = \bigcup_j E_j$ be a countable disjoint union of s-sets with $\mathscr{H}^s(E) < \infty$. Then for any k,

$$\underline{D}^s(E_k, x) = \underline{D}^s(E, x) \text{ and } \bar{D}^s(E_k, x) = \bar{D}^s(E, x).$$

at almost all $x \in E_k$.

Proof. Apply Corollary 2.6 to the subset E_k of E.   □

Corollary 2.8
Let E be an s-set. If E is regular resp, irregular then any measurable subset of E of positive measure is regular resp. irregular.

Proof. This is immediate from Corollary 2.6 and the definition of regularity.   □

Corollary 2.9
The intersection of a regular resp. irregular set and a measurable set is a regular resp. irregular set. The intersection of a regular set and an irregular set is of measure zero.

Proof. The first pair of statements follow from Corollary 2.8. The final statement holds since such an intersection must be both regular and irregular.   □

The next corollary enables us to treat the regular and irregular parts of an s-set independently.

Corollary 2.10 (decomposition theorem)
If E is an s-set, the set of regular points of E is a regular set, and the set of irregular points of E is an irregular set.

Proof. By Lemma 2.1 the sets of regular and irregular points are both measurable, so this corollary follows from Corollary 2.6.   □

We can also obtain bounds for the upper *angular* densities of s-sets in $\mathbb{R}^n(n \geq 2)$. Angular densities were introduced by Besicovitch (1929, 1938) to study tangential properties of s-sets. If θ is a unit vector and ϕ an angle, let $S(x, \theta, \phi)$ be the closed one-way infinite cone with vertex x and axis in direction θ consisting of those points y for which the segment $[x, y]$ makes an angle of, at most, ϕ with θ. Write $S_r(x, \theta, \phi) = B_r(x) \cap S(x, \theta, \phi)$ for the corresponding spherical sector of radius r. Angular densities are defined analogously to spherical densities, but with $S_r(x, \theta, \phi)$ replacing $B_r(x)$. Thus

$$\bar{D}^s(E, x, \theta, \phi) = \overline{\lim_{r \to 0}} \frac{\mathscr{H}^s(E \cap S_r(x, \theta, \phi))}{(2r)^s}$$

and

$$\underline{D}^s(E, x, \theta, \phi) = \underline{\lim_{r \to 0}} \frac{\mathscr{H}^s(E \cap S_r(x, \theta, \phi))}{(2r)^s}$$

are the *upper* and *lower angular densities of E* at x. A routine check (see Lemma 2.1) establishes all the desirable measurability properties for the angular densities and the sets associated with them.

Comparing the diameter of a spherical sector with its radius, it follows from the definitions that

$$\bar{D}^s(E, x, \theta, \phi) \leq 2^{-s}|S_1(x, \theta, \phi)|^s \bar{D}^s_c(E, x).$$

We deduce immediately from Theorem 2.3 that, given θ and ϕ,

$$\bar{D}^s(E, x, \theta, \phi) \leq 2^{-s}|S_1(x, \theta, \phi)|^s$$

for almost all $x \in E$. On calculating the diameters of the sectors, this becomes

$$\bar{D}^s(E, x, \theta, \phi) \leq \begin{cases} 2^{-s} & (0 < \phi \leq \frac{1}{6}\pi) \\ (\sin \phi)^s & (\frac{1}{6}\pi \leq \phi \leq \frac{1}{2}\pi) \\ 1 & (\frac{1}{2}\pi \leq \phi \leq \pi). \end{cases}$$

We may obtain a positive lower bound for the upper angular densities if $\phi \geq \frac{1}{2}\pi$ by modifying part (a) of the proof of Theorem 2.3, see Exercise 2.3. Then

$$\bar{D}^s(E, x, \theta, \phi) \geq \begin{cases} 0 & (0 < \phi < \frac{1}{2}\pi) \\ 2^{-s} & (\frac{1}{2}\pi \leq \phi \leq \pi). \end{cases}$$

Estimates for lower densities are somewhat harder to derive, and will be considered later in certain cases in connection with tangency properties.

Morgan (1935), Gills (1935), Besicovitch (1938) and Dickinson (1939) give some constructions of sets for which the densities and angular densities take extreme values.

Exercises on Chapter 2

2.1   If E is the Cantor set, show that $\underline{D}^s(E, x) \leq 2^{-s}$ for all x, where $s = \log 2/\log 3$. Deduce that E is irregular.

2.2   Define the upper cubical density at x of an s-set E in $\mathbb{R}^n$ as $\overline{\lim}_{r \to \infty} \mathscr{H}^s(E \cap S_r(x))/|S_r(x)|^s$, where here $S_r(x)$ denotes the cube of side r centred at x with sides parallel to the coordinate axes. Show that at almost all $x \in E$ the upper cubical density of E lies between $2^{-s}n^{-s/2}$ and 1.

2.3   Show that if E is an s-set in $\mathbb{R}^n (n \geq 2)$ and θ is a unit vector, then $2^{-s} \leq \bar{D}^s(E, x, \theta, \frac{1}{2}\pi)$ for almost all $x \in E$. (Follow the proof of Theorem 2.3(a); you will need to use the regularity of $\mathscr{H}^s$ to obtain a closed set to work with.) Deduce that $2^{-s} \leq \bar{D}^s(E, x, \theta, \phi)$ if $\phi \geq \frac{1}{2}\pi$.

2.4   Let $\psi : \mathbb{R}^n \to \mathbb{R}^n$ be a continuously differentiable transformation with non-vanishing Jacobian. If E is an s-set show that $\bar{D}^s(\psi(E), \psi(x)) = \bar{D}^s(E, x)$ for all x, with a similar result for lower densities.

2.5   Use Corollary 2.4 and Theorem 1.12 to prove the Lebesgue density theorem, Theorem 1.13.

3
Structure of sets of integral dimension

3.1 Introduction

In this chapter we discuss the density and tangency structure of *s*-sets in $\mathbb{R}^n$ when *s* is an integer. We know from Corollary 2.10 that an *s*-set splits into a regular part and an irregular part, and we find that these two types of set exhibit markedly different properties. One of our aims is to characterize regular sets as subsets of countable unions of rectifiable curves or surfaces, and thus to relate the measure theoretic and the descriptive topological ideas.

We present in detail the theory of linearly measurable sets or 1-sets in $\mathbb{R}^2$. This work is almost entirely due to Besicovitch (1928a, 1938), the latter paper including some improved proofs as well as further results. Most of his proofs seem hard to better except in relatively minor ways and, hopefully, in presentation. Certainly, some of the geometrical methods used by Besicovitch involve such a degree of ingenuity that it is surprising that they were ever thought of at all. Some of the work in this chapter is also described in de Guzmán (1981).

3.2 Curves and continua

Regular 1-sets and rectifiable curves are intimately related. Indeed, a regular 1-set is, to within a set of measure zero, a subset of a countable collection of rectifiable curves. This section is devoted to a study of curves, mainly from a topological viewpoint and in relation to continua of finite linear measure. Here we work in $\mathbb{R}^n$ as the theory is no more complicated than for plane curves.

A *curve* (or *Jordan curve*) Γ is the image of a continuous injection $\psi : [a, b] \to \mathbb{R}^n$, where $[a, b] \subset \mathbb{R}$ is a closed interval. Any curve is a *continuum*, that is, a compact connected set. This follows since the continuous image of any compact connected set is compact and connected. In particular, any curve is a Borel set and so is $\mathcal{H}^s$-measurable. Moreover, a continuous bijection between compact sets has a continuous inverse, so that a curve may be defined as a homeomorphic image of a closed interval.

The *length* of the curve Γ is defined as

$$\mathcal{L}(\Gamma) = \sup \sum_{i=1}^{m} |\psi(t_i) - \psi(t_{i-1})| \tag{3.1}$$

where the supremum is taken over all dissections $a = t_0 < t_1 < \ldots < t_m = b$ of $[a, b]$. If $\mathscr{L}(\Gamma) < \infty$ (that is, if ψ is of bounded variation), then Γ is said to be *rectifiable*.

Note that our definition excludes self-intersecting curves, which are covered by the following lemma.

Lemma 3.1

Let $\psi : [a, b] \to \mathbb{R}^n$ be a continuous mapping, with $\psi(a) \neq \psi(b)$. Then $\psi[a, b]$ contains a curve joining $\psi(a)$ to $\psi(b)$.

Proof. For each multiple point x of $[a, b]$ let I_x be the largest (closed) interval $[t_1, t_2]$ with $\psi(t_1) = \psi(t_2) = x$. Let $\mathscr{I}$ denote the collection of such intervals that are contained in no others. Then $\mathscr{I}$ consists of countably many disjoint proper closed intervals. Thus we may construct a continuous surjection $f : [a, b] \to [0, 1]$ such that $f(a) = 0$, $f(b) = 1$, and such that if $t_1 \leq t_2$, then $f(t_1) \leq f(t_2)$ with equality if and only if t_1 and t_2 lie in a common interval of $\mathscr{I}$. Define $\psi_0 : [0, 1] \to \mathbb{R}^n$ by $\psi_0(u) = x$ if $f^{-1}(u) = I_x$ for some $I_x \in \mathscr{I}$ and by $\psi_0(u) = \psi(f^{-1}(u))$ otherwise. It is easy to check that ψ_0 is a continuous injection with $\psi_0(0) = \psi(a)$ and $\psi_0(1) = \psi(b)$. $\quad\square$

The sum in (3.1) does not decrease with the introduction of additional dissection points. It is clear that if $\psi[a, b]$ is split into two curves $\psi[a, c]$ and $\psi[c, b]$ where $a < c < b$, then the sum of the lengths of the new curves equals the length of the original curve, i.e. $\mathscr{L}$ is additive on curves with a common endpoint.

It is always possible to parametrize a rectifiable curve Γ *by arc length*, that is, to represent Γ as the image of a function $\psi_0 : [0, \mathscr{L}(\Gamma)] \to \mathbb{R}$ in such a way that the length of $\psi_0[0, t]$ is t. This may be achieved by letting $\psi_0(t)$ be the unique point $\psi(u)$ for which $\mathscr{L}(\psi[a, u]) = t$. If ψ represents the rectifiable curve Γ by arc length we see from (3.1) that

$$|\psi(t_1) - \psi(t_2)| \leq |t_1 - t_2|. \tag{3.2}$$

In particular, this implies that ψ is an absolutely continuous function.

For a further discussion on the definition of curves see Burkill & Burkill (1970, p. 246) or Pelling (1977).

Lemma 3.2

If Γ is a curve, then $\mathscr{H}^1(\Gamma) = \mathscr{L}(\Gamma)$.

Proof. Let Γ be a curve joining z and w. If proj denotes orthogonal projection from $\mathbb{R}^n$ onto the straight line through z and w, then $|\text{proj } x - \text{proj } y| \leq |x - y|$ if $x, y \in \mathbb{R}^n$. By Lemma 1.8 $\mathscr{H}^1(\Gamma) \geq \mathscr{H}^1(\text{proj } \Gamma) \geq \mathscr{H}^1([z, w]) = \mathscr{L}^1([z, w]) = |z - w|$, since proj $\Gamma \supset [z, w]$.

Now suppose that Γ is defined by $\psi:[a,b] \to \mathbb{R}^n$. By the above remark, $\mathscr{H}^1(\psi[t,u]) \geq |\psi(t) - \psi(u)|$ for any t and u. Then if $a = t_0 < t_1 < \ldots < t_m = b$ is any dissection of $[a,b]$,

$$\sum_i |\psi(t_i) - \psi(t_{i-1})| \leq \sum_i \mathscr{H}^1(\psi[t_{i-1}, t_i]) = \mathscr{H}^1(\Gamma),$$

since the arcs $\psi[t_{i-1}, t_i]$ of Γ are disjoint except for endpoints. Thus $\mathscr{L}(\Gamma) \leq \mathscr{H}^1(\Gamma)$.

Finally, assume that $\mathscr{L}(\Gamma) < \infty$ and let ψ parametrize Γ by arc length. Since ψ is a surjection from $[0, \mathscr{L}(\Gamma)]$ to Γ with (3.2) holding, Lemma 1.8 implies that $\mathscr{H}^1(\Gamma) \leq \mathscr{H}^1([0, \mathscr{L}(\Gamma)]) = \mathscr{L}(\Gamma)$. $\square$

One consequence of this lemma is that if Γ is a rectifiable curve, then $\mathscr{H}^s(\Gamma)$ is infinite if $s < 1$ and zero if $s > 1$ (see (1.14)).

Corollary 3.3

Let ψ parametrize the rectifiable curve Γ by arc length. If E is any Lebesgue-measurable subset of $[0, \mathscr{L}(\Gamma)]$, then $\psi(E)$ is an $\mathscr{H}^1$-measurable subset of $\mathbb{R}^n$ and $\mathscr{H}^1(\psi(E)) = \mathscr{L}^1(E)$.

Proof. By Lemma 3.2 the additive set functions $\mathscr{H}^1(\psi(.))$ and $\mathscr{L}^1(.)$ agree on closed intervals, so by the usual process of extension of measures (see, for example, Kingman & Taylor (1966)) they agree on the Lebesgue-measurable subsets of $[0, \mathscr{L}(\Gamma)]$. $\square$

This corollary allows all the usual results on Lebesgue measure on the line to be transferred to curves. For example, it follows from the Lebesgue density theorem, Theorem 1.13, that if E is an $\mathscr{H}^1$-measurable subset of a curve Γ, then at almost all $x \in E$ we have $\mathscr{H}^1(E \cap I)/\mathscr{H}^1(I) \to 1$ as $\mathscr{H}^1(I) \to 0$, where I is a subarc of Γ containing x.

We do not make further use of this corollary.

We have frequent recourse to the next lemma:

Lemma 3.4

Let E be a continuum containing x and y. If $|x - y| = \rho$, then $\mathscr{H}^1(E \cap B_\rho(x)) \geq \rho$. In particular, $\mathscr{H}^1(E) \geq |E|$.

Proof. Let $f : \mathbb{R}^n \to [0, \infty)$ be defined by $f(z) = |z - x|$. Then f is a continuous mapping such that

$$|f(z) - f(w)| \leq |z - w| \qquad (z, w \in \mathbb{R}^n). \tag{3.3}$$

The set $f(E \cap B_\rho(x))$ contains the interval $[0, \rho]$, otherwise, for some $r (0 < r < \rho), E = (E \cap B_r(x)) \cup (E \setminus B_r(x))$ would be a decomposition of E into disjoint closed sets. Applying Lemma 1.8 to the mapping f

$$\mathcal{H}^1(E \cap B_\rho(x)) \geq \mathcal{H}^1(f(E \cap B_\rho(x)))$$
$$\geq \mathcal{H}^1([0,\rho]) = \mathcal{L}^1([0,\rho]) = \rho. \quad \square$$

We may now deduce the basic density property of rectifiable curves.

Lemma 3.5

A rectifiable curve is a regular 1-set.

Proof. If Γ is rectifiable, $\mathcal{L}(\Gamma) < \infty$, so by Lemma 3.2, $\mathcal{H}^1(\Gamma) < \infty$. As Γ contains at least two distinct points, Lemma 3.4 implies $\mathcal{H}^1(\Gamma) > 0$, so Γ is an 1-set.

Let x be a point of Γ other than an endpoint, and suppose x divides Γ into two rectifiable subcurves Γ_- and Γ_+. By Lemma 3.4, $\mathcal{H}^1(\Gamma_- \cap B_\rho(x)) \geq \rho$ and $\mathcal{H}^1(\Gamma_+ \cap B_\rho(x)) \geq \rho$ if ρ is sufficiently small, giving $\mathcal{H}^1(\Gamma \cap B_\rho(x)) \geq 2\rho$. Thus at all points of Γ other than the endpoints, $D^1(\Gamma, x) \geq 1$. Taken together with Corollary 2.5 this implies that $D^1(\Gamma, x)$ exists and equals 1 at almost all $x \in \Gamma$. $\quad \square$

Next we discuss the existence of tangents to rectifiable curves. We say that an s-set E in $\mathbb{R}^n$ has a *tangent* at x in the direction $\pm \boldsymbol{\theta}$ if $\bar{D}^s(E, x) > 0$ and for every angle $\phi > 0$,

$$\lim_{r \to 0} r^{-s} \mathcal{H}^s(E \cap (B_r(x) \backslash S_r(x, \boldsymbol{\theta}, \phi) \backslash S_r(x, -\boldsymbol{\theta}, \phi))) = 0 \qquad (3.4)$$

(the line through x in the direction $\pm \boldsymbol{\theta}$ is of course the tangent line). Clearly, an s-set can have, at most, one tangent at each of its points.

We give Besicovitch's (1944) elementary proof that a rectifiable curve has a tangent at almost all of its points. An alternative approach would be to appeal to standard results on the differentiability of functions of bounded variation, see Kingman & Taylor (1966, Section 9.1).

Lemma 3.6

Let Γ be a rectifiable curve with endpoints x and y, and let ϕ be a positive angle. Let E be the set of points on Γ that belong to pairs of arbitrarily small subarcs of Γ subtending chords that make an angle of more than 2ϕ with each other. Then $\mathcal{H}^1(E) \leq (\mathcal{L}(\Gamma) - |x - y|)/(1 - \cos \phi)$.

Proof. Let L denote the line through x and y and let $\mathcal{V}$ be the collection of closed subarcs of Γ subtending chords that make angles of more than ϕ with L. By the conditions of the lemma, $\mathcal{V}$ is a Vitali class for E. Hence using the covering theorem, Theorem 1.10(b), we may, for any $\varepsilon > 0$, find a finite collection $\Gamma_1, \ldots, \Gamma_m$ of disjoint subarcs of Γ belonging to $\mathscr{V}$ such that

$$\mathscr{H}^1(E) \leq \sum_{i=1}^{m} |\Gamma_i| + \varepsilon \leq \sum_{i=1}^{m} \mathscr{L}(\Gamma_i) + \varepsilon. \tag{3.5}$$

If $\Gamma_0', \Gamma_1', \ldots, \Gamma_m'$ are the (possibly null) complementary arcs, then projecting orthogonally onto L and using the fact that projection does not increase length, we see that

$$\cos \phi \sum_i \mathscr{L}(\Gamma_i) + \sum_i \mathscr{L}(\Gamma_i') \geq |x - y|.$$

Since $\sum_i \mathscr{L}(\Gamma_i) + \sum_i \mathscr{L}(\Gamma_i') = \mathscr{L}(\Gamma)$ we get, using (3.5),

$$\mathscr{H}^1(E) - \varepsilon \leq \sum_i \mathscr{L}(\Gamma_i) \leq (\mathscr{L}(\Gamma) - |x - y|)/(1 - \cos \phi).$$

The observation that ε is arbitrary completes the proof. $\qquad \square$

Corollary 3.7

If $\phi > 0$ and E is the set of points on a rectifiable curve Γ that belong to pairs of arbitrarily small subarcs of Γ subtending chords that make an angle of more than 2ϕ with each other, then $\mathscr{H}^1(E) = 0$.

Proof. Given $\varepsilon > 0$ we may by (3.1) find points $x_0, x_1, \ldots, x_m$ on Γ (in that order and with x_0 and x_m the endpoints of Γ) such that

$$\mathscr{L}(\Gamma) < \sum_{i=1}^{m} |x_i - x_{i-1}| + \varepsilon.$$

Write Γ_i for the portion of Γ between x_{i-1} and x_i. Then applying Lemma 3.6 to each Γ_i in turn gives

$$\mathscr{H}^1(E) = \sum_{i=1}^{m} \mathscr{H}^1(E \cap \Gamma_i) \leq \sum_{i=1}^{m} (\mathscr{L}(\Gamma_i) - |x_i - x_{i-1}|)/(1 - \cos \phi)$$

$$= (\mathscr{L}(\Gamma) - \sum_{i=1}^{m} |x_i - x_{i-1}|)/(1 - \cos \phi) < \varepsilon/(1 - \cos \phi).$$

Thus $\mathscr{H}^1(E) = 0$. $\qquad \square$

Theorem 3.8

A rectifiable curve Γ has a tangent at almost all of its points.

Proof. Since the continuum Γ has at least two points, Lemma 3.4 implies that $\bar{D}^s(E, x) \geq \frac{1}{2}$ at all $x \in \Gamma$.

Let $\psi : [a, b] \to \mathbb{R}^n$ be a defining function for Γ. It follows from Corollary 3.7 that for almost all $x \in \Gamma$ we may find a unit vector θ such that, given $\phi > 0$,

$$\psi(u) \in S(x, \theta, \phi) \cup S(x, -\theta, \phi)$$

if $|u-t|<\varepsilon$, where $\phi(t)=x$. Moreover, if ρ is positive, $\psi(u)\notin B_\rho(x)$ if $|u-t|\geq\varepsilon$ (otherwise choose a sequence of numbers $\{u_i\}$ with $|u_i-t|\geq\varepsilon$ and $\psi(u_i)\to x$; then using sequential compactness there exists u with $|u-t|\geq\varepsilon$ and $\psi(u)=x=\psi(t)$ giving a double point of Γ). Thus

$$\Gamma\cap(B_\rho(x)\backslash S_\rho(x,\boldsymbol{\theta},\phi)\backslash S_\rho(x,-\boldsymbol{\theta},\phi))=\varnothing,$$

with (3.4) as a trivial consequence. $\quad\square$

Tangents to curves are often thought of in terms of differentiability. A minor variant of the proof of Theorem 3.8 shows that a rectifiable curve defined by a function ψ has a tangent at $\psi(t)$ provided the derivative $\psi'(t)$ exists and is non-vanishing.

Besicovitch (1934a, 1956, 1957, 1960) also considers how the definition of a tangent can be adapted for curves and sets of infinite linear measure.

A 1-set contained in a countable union of rectifiable curves is called a *Y-set*.

Lemmas 3.5 and 3.8 have counterparts for *Y*-sets.

Corollary 3.9
A Y-set is a regular 1-set.

Proof. A 1-set contained in a rectifiable curve has density 1 almost everywhere, applying Corollary 2.8 to Lemma 3.5. The same is true of a *Y*-set by Corollary 2.7. $\quad\square$

Corollary 3.10
A Y-set has a tangent at almost all of its points.

Proof. By Corollary 3.9, a *Y*-set has positive lower density almost everywhere. By Corollary 2.6 applied to Theorem 3.8 a 1-set contained in a rectifiable curve satisfies (3.4) almost everywhere, so by Corollary 2.7 a *Y*-set has a tangent at almost all of its points. $\quad\square$

Eventually we shall show that a regular 1-set is a *Y*-set, together with a set of measure zero, and it will follow from Corollary 3.10 that any regular 1-set has a tangent almost everywhere.

A set with the property described in the next lemma is a sometimes called a *Lipschitz set*.

Lemma 3.11
Let E be a bounded subset of $\mathbb{R}^n$ such that if $x,y\in E$ the segment $[x,y]$ makes an angle of, at most, $\phi<\frac{1}{2}\pi$ with a fixed line L. Then E is a subset of a rectifiable curve.

Proof. We may assume that E is closed. Let $\Pi(t)$ be the hyperplane perpendicular to L at distance t from some origin, and let a and b be the extreme values of t for which $\Pi(t)$ intersects E. Then $\Pi(t)$ can contain, at most, one point of E for each t; let $\psi(t)$ denote this point if it exists, otherwise if $a < t < b$ let $\psi(t)$ be the point of $\Pi(t)$ on the line segment joining the points of E nearest to $\Pi(t)$ on either side. The segment $[\psi(t_1), \psi(t_2)]$ makes an angle of, at most, ϕ with L if $t_1 \neq t_2$, so $|\psi(t_1) - \psi(t_2)| \leq |t_1 - t_2|/\cos\phi$. By (3.1), ψ defines a rectifiable curve.   □

Next we show that a continuum of finite $\mathcal{H}^1$-measure is, to within a set of measure zero, a countable collection of rectifiable curves and is therefore regular. Following Besicovitch (1938) we use arcwise connectivity as an intermediate stage. Recall that a set E is *arcwise connected* if, given $z, w \in E$, there is a curve lying inside E that joins z and w. A compact arcwise connected set is easily seen to be a continuum. We require the following converse.

Lemma 3.12

A continuum E with $\mathcal{H}^1(E) < \infty$ is arcwise connected.

Proof. This is essentially a consequence of Lemma 3.4. Take $z, w \in E$. As E is connected, E is *chain connected*, so that for each $\varepsilon > 0$ we may find a chain of points $z = x_0, x_1, \ldots, x_m = w$ in E with $|x_i - x_{i-1}| \leq \varepsilon$ for $1 \leq i \leq m$. (This follows by showing that the set of points chain connected to z is both open and closed in E.) By deleting intermediate points of the chain if necessary we may assume that $|x_i - x_j| > \varepsilon$ if $|i - j| \geq 2$. Thus no point of $\mathbb{R}^n$ lies in more than two of the balls $B_{\frac{1}{2}\varepsilon}(x_i)$, so, assuming that $m \geq 2$ and using Lemma 3.4,

$$2\mathcal{H}^1(E) \geq \sum_{i=0}^{m} \mathcal{H}^1(E \cap B_{\frac{1}{2}\varepsilon}(x_i)) \geq \tfrac{1}{2}m\varepsilon.$$

Hence if Γ_ε denotes the polygonal curve (necessarily not self-intersecting) obtained by joining $x_0, x_1, \ldots, x_m$,

$$\mathscr{L}(\Gamma_\varepsilon) = \sum_{i=1}^{m} |x_i - x_{i-1}| \leq m\varepsilon \leq 4\mathcal{H}^1(E). \tag{3.6}$$

For $0 \leq t \leq 1$ let $\psi_\varepsilon(t)$ be the point on Γ_ε a proportion t along its length, so that the part of Γ_ε lying between $z = \psi_\varepsilon(0)$ and $\psi_\varepsilon(t)$ has length $t\mathscr{L}(\Gamma_\varepsilon)$. Then if $0 \leq t_1 < t_2 \leq 1$,

$$|\psi_\varepsilon(t_1) - \psi_\varepsilon(t_2)| \leq \mathscr{L}(\psi_\varepsilon[t_1, t_2]) \leq |t_2 - t_1| 4\mathcal{H}^1(E), \tag{3.7}$$

by (3.6). The curves $\{\Gamma_\varepsilon\}_{\varepsilon \leq 1}$ all lie in a bounded subset of $\mathbb{R}^n$, so the family of functions $\{\psi_\varepsilon\}_{\varepsilon \leq 1}$ from $[0, 1]$ to $\mathbb{R}^n$ is uniformly bounded and, by (3.7), is equicontinuous. By the Arzelà–Ascoli theorem (see, for example, Dunford &

Schwartz (1958, Section IV 6.7)), such a family is sequentially compact, that is, there is a sequence $\varepsilon(j) \to 0$ and a continuous function $\psi:[0,1] \to \mathbb{R}^n$ such that $\psi_{\varepsilon(j)}$ converges to ψ uniformly on $[0, 1]$. Let $\psi[0, 1] = \Gamma$. Clearly, $\psi(0) = z$ and $\psi(1) = w$. Also, if $x \in \Gamma$, then, given $\delta > 0$, we may find $\varepsilon(j) < \frac{1}{2}\delta$ such that $\Gamma_{\varepsilon(j)}$ meets the ball $B_{\frac{1}{2}\delta}(x)$. Thus there is some point of E (that is, a vertex of $\Gamma_{\varepsilon(j)}$) in $B_\delta(x)$. As E is closed we conclude that $x \in E$, and so $\Gamma \subset E$. If ψ is injective, then Γ is a curve. Otherwise we may reduce Γ to a curve joining z and w by Lemma 3.1.

Lemma 3.13

Any compact arcwise connected set E with $\mathcal{H}^1(E) < \infty$ consists of a countable union of rectifiable curves, together with a set of $\mathcal{H}^1$-measure zero.

Proof. We define a sequence of curves $\{\Gamma_j\}$ inductively. Let Γ_1 be a curve in E joining two of the most distant points of E. Suppose that the curves $\Gamma_1, \Gamma_2, \ldots, \Gamma_k$ have been defined. Let x be a point of E at maximum distance, d_k say, from $\bigcup_1^k \Gamma_j$ (this maximum is attained as the sets involved are compact). If $d_k = 0$ the process terminates and the result follows. Otherwise let Γ_{k+1} be a curve in E joining x to $\bigcup_1^k \Gamma_j$ with Γ_{k+1} disjoint from $\bigcup_1^k \Gamma_j$ except for an endpoint. By Lemma 3.4,

$$\mathcal{H}^1(\Gamma_{k+1}) \geq d_k. \tag{3.8}$$

Then

$$\sum_{j=1}^\infty d_j \leq \sum_1^\infty \mathcal{H}^1(\Gamma_j) \leq \mathcal{H}^1(E) < \infty, \tag{3.9}$$

so that $d_j \to 0$. If $x \in E$ and the distance d from x to $\bigcup_1^\infty \Gamma_j$ is positive, then $d_j < d$ for some j, and x would have been chosen as the endpoint of Γ_{j+1}, a contradiction. We conclude that E is the closure of $\bigcup_1^\infty \Gamma_j$. By Lemma 3.2, $\mathcal{L}(\Gamma_j) = \mathcal{H}^1(\Gamma_j) \leq \mathcal{H}^1(E) < \infty$, so that Γ_j is a rectifiable curve for each j.

Finally, for each value of k, $\bigcup_1^k \Gamma_j$ is closed, so

$$\mathcal{V} = \left\{ B_r(x) : x \in E \backslash \bigcup_1^k \Gamma_j \quad \text{and } B_r(x) \cap \bigcup_1^k \Gamma_j = \varnothing \right\}$$

is a Vitali class of balls for $E \backslash \bigcup_1^k \Gamma_j$. Let x be the centre of some $B \in \mathcal{V}$. Since x lies in the closure of $\bigcup_{k+1}^\infty \Gamma_j$ there are points of this union arbitrarily close to x. Such points must be connected via a sequence of arcs in $\bigcup_{k+1}^\infty \Gamma_j$ to a point of $\bigcup_1^k \Gamma_j$ necessarily outside B, so, by Lemma 3.4,

$$\tfrac{1}{2}|B| \leq \mathcal{H}^1\left(B \cap \bigcup_{k+1}^\infty \Gamma_j \right)$$

for any $B \in \mathcal{V}$. Thus, given $\varepsilon > 0$, we may use the Vitali theorem, Theorem

1.10(b), to choose a disjoint sequence of balls $\{B_i\}$ from $\mathscr{V}$ such that

$$\mathscr{H}^1\left(E\backslash\bigcup_1^k \Gamma_j\right) \le \sum_i |B_i| + \varepsilon \le 2\sum_i \mathscr{H}^1\left(B_i \cap \bigcup_{k+1}^\infty \Gamma_j\right) + \varepsilon$$

$$\le 2\mathscr{H}^1\left(\bigcup_{k+1}^\infty \Gamma_j\right) + \varepsilon.$$

Hence

$$\mathscr{H}^1\left(E\backslash\bigcup_1^k \Gamma_j\right) \le 2\mathscr{H}^1\left(\bigcup_{k+1}^\infty \Gamma_j\right).$$

By (3.9) the right-hand side of this inequality tends to zero as $k \to \infty$ so that $\mathscr{H}^1(E\backslash\bigcup_1^\infty \Gamma_j) = 0$, as required. $\square$

The above lemma appears to have been proved by Ważewski (1927) independently of Besicovitch.

Theorem 3.14
Let E be a continuum with $\mathscr{H}^1(E) < \infty$. Then E consists of a countable union of rectifiable curves, together with a set of $\mathscr{H}^1$-measure zero.

Proof. Combine Lemmas 3.12 and 3.13. $\square$

Corollary 3.15
Let E be a continuum with $\mathscr{H}^1(E) < \infty$. Then E is regular and has a tangent at almost all of its points.

Proof. By Theorem 3.14, E is a Y-set, together with a set of measure zero, so the result is immediate from Corollaries 3.9 and 3.10. (A set of measure zero does not affect the value of the densities and angular densities.) $\square$

Several other results on enclosing regular 1-sets in curves have been formulated. For example, the exceptional set of points in Theorem 3.14 may be regarded as a subset of length 0 of a possibly self-intersecting rectifiable curve. Similarly, any continuum E of finite measure may be enclosed in a (self-intersecting) rectifiable curve of length, at most, $2\mathscr{H}^1(E)$ (see Exercise 3.5).

We complete this section by proving a theorem originally due to Gołąb on the semicontinuity of the linear measure of continua. This result is of considerable importance in geometric measure theory and is used in geometrical problems to assert the existence of continua or curves of maximal measure with specified properties. We employ the well-known selection theorem of Blaschke (1916); this theorem is discussed in a wider setting in Rogers (1970).

If $E \subset \mathbb{R}^n$, the *δ-parallel body* of E is the closed set of points within distance δ of E, that is,

$$[E]_\delta = \{x \in \mathbb{R}^n : \inf_{y \in E} |x - y| \leq \delta\}.$$

The *Hausdorff metric* δ is defined on the collection of all non-empty compact subsets of $\mathbb{R}^n$ by

$$\delta(E, F) = \inf\{\delta : E \subset [F]_\delta \text{ and } F \subset [E]_\delta\}.$$

It is a simple exercise to show that δ *is* a metric.

Theorem 3.16 (Blaschke selection theorem)
Let $\mathscr{C}$ be an infinite collection of non-empty compact sets all lying in a bounded portion B of $\mathbb{R}^n$. Then there exists a sequence $\{E_j\}$ of distinct sets of $\mathscr{C}$ convergent in the Hausdorff metric to a non-empty compact set E.

Proof. First we produce a Cauchy sequence of sets from $\mathscr{C}$. Let $\{E_{1,i}\}_i$ be any sequence of distinct sets of $\mathscr{C}$. For each $k > 1$ we define an infinite subsequence $\{E_{k,i}\}_i$ of $\{E_{k-1,i}\}_i$ as follows. Let $\mathscr{B}_k$ be a finite collection of closed balls of diameter, at most, $1/k$ covering B. Each $E_{k-1,i}$ intersects some specific combination of these balls so there must be an infinite subcollection $\{E_{k,i}\}_i$ of $\{E_{k-1,i}\}_i$ which all intersect precisely the same balls of $\mathscr{B}_k$, by the pigeon-hole principle (only a finite number of different combinations are available). If F is the union of the balls of $\mathscr{B}_k$ in this particular combination, then $E_{k,i} \subset F \subset [E_{k,i}]_{1/k}$ for all i so that $\delta(E_{k,i}, F) \leq 1/k$, giving $\delta(E_{k,i}, E_{k,j}) \leq 2/k$ for all i, j. Letting $E_i = E_{i,i}$ we see at once that

$$\delta(E_i, E_j) \leq 2/\min\{i, j\}, \tag{3.10}$$

so $\{E_i\}_i$ is a Cauchy sequence.

Now

$$E = \bigcap_{j=1}^{\infty} \overline{\bigcup_{i=j}^{\infty} E_i}$$

is a non-empty compact set, being the intersection of a decreasing sequence of non-empty compact sets. (A bar denotes closure.) By (3.10), $\overline{\bigcup_{i=j}^{\infty} E_i} \subset [E_j]_{2/j}$, so $E \subset [E_j]_{2/j}$ for all j. On the other hand, if $x \in E_j$, then by (3.10) $x \in [E_i]_{2/j}$ if $i \geq j$, so $x \in [\overline{\bigcup_{i=k}^{\infty} E_i}]_{2/j}$ if $k \geq j$. Choose $y_k \in \overline{\bigcup_{i=k}^{\infty} E_i}$ with $|x - y_k| \leq 2/j$. By sequential compactness a subsequence of $\{y_k\}$ converges to some $y \in \mathbb{R}^n$ with $|x - y| \leq 2/j$. But $y \in \bigcap_{k=1}^{\infty} \overline{\bigcup_{i=k}^{\infty} E_i} = E$, so $x \in [E]_{2/j}$. We conclude that $E_j \subset [E]_{2/j}$ and hence that $\delta(E, E_j) \leq 2/j$. Thus $\{E_j\}$ converges to E in the Hausdorff metric. $\square$

Notice that the latter half of the above proof is essentially a demonstration that δ is a *complete* metric.

A continuum in which every pair of points is joined by a unique rectifiable path is called a *tree*. The next lemma allows us to prove the semicontinuity theorem for a sequence of trees, and we extend the result to general continua by approximation.

Lemma 3.17

Let $F \subset \mathbb{R}^n$ be a tree with $\mathcal{H}^1(F) < \infty$. Then, given $\delta > 0$, we can write $F = \bigcup_{i=1}^{k} F_i$, where the F_i are continua such that

(a) $\sum_{i=1}^{k} |F_i| \leq \sum_{i=1}^{k} \mathcal{H}^1(F_i) = \mathcal{H}^1(F)$,

(b) $|F_i| \leq \delta$ *for all* i,

(c) $k \leq 3\delta^{-1} \mathcal{H}^1(F) + 1$.

Proof. If $|F| \leq \delta$ we may take $k = 1$ and $F_1 = F$, so assume that $|F| > \delta$. Let x be any point of F, and let m denote the supremum of the path-distances of points of F from x; then by Lemma 3.4 $m \leq \mathcal{H}^1(F) < \infty$, while $m > \frac{1}{2}\delta$ because, otherwise, $|F| \leq \delta$. Let y be a point of F at path-distance greater than $m - \frac{1}{6}\delta$ from x, and let z be the point on the unique path joining y to x at path-distance $m - \frac{1}{2}\delta$ from x. The point z determines a dissection of F into two subtrees F_1 and F' with z as their only common point, where F_1 consists of those points of F whose joins to x pass through z, see Figure 3.1. Every point of F_1 is within path-distance $\frac{1}{2}\delta$ of z, so $|F_1| \leq \delta$. By Lemma 3.4, $|F_1| \leq \mathcal{H}^1(F_1)$, and also $\mathcal{H}^1(F_1)$ is greater than the path-distance from y to z, which is at least $(m - \frac{1}{6}\delta) - (m - \frac{1}{2}\delta) = \frac{1}{3}\delta$.

If $|F'| > \delta$ we repeat this process with the tree F' to break off a subtree F_2, and so on, until we are left with a tree F_k of diameter, at most, δ. Parts (a) and (b) of the conclusion are immediate, while (c) follows from (a) since $\frac{1}{3}\delta \leq \mathcal{H}^1(F_i)$ if $1 \leq i \leq k - 1$ $\square$

Fig. 3.1

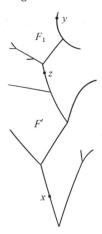

Theorem 3.18

Let $\{E_j\}$ be a sequence of continua in $\mathbb{R}^n$ convergent in the Hausdorff metric to a compact set E. Then E is a continuum and

$$\mathcal{H}^1(E) \leq \varliminf_{j \to \infty} \mathcal{H}^1(E_j).$$

Proof. If E is not connected, then there is a disjoint decomposition $E = G_1 \cup G_2$, where G_1 and G_2 are non-empty closed sets, with the distance δ between G_1 and G_2 strictly positive. Then for j large enough, $E_j \subset [G_1]_{\frac{1}{4}\delta} \cup [G_2]_{\frac{1}{4}\delta}$, where $[G_1]_{\frac{1}{4}\delta}$ and $[G_2]_{\frac{1}{4}\delta}$ are disjoint parallel bodies containing points of E_j, and this contradicts the connectedness of E_j. Thus E is a continuum.

To prove the inequality we may assume that $\mathcal{H}^1(E_j) \leq c < \infty$ for all j. For each j choose a finite subset S_j of E_j in such a way that S_j also converges to E in the Hausdorff metric. (Let $\delta_j \to 0$, and take a finite set S_j such that every point of E_j lies within δ_j of some point in S_j.) By Lemma 3.12 the sets E_j are arcwise connected, so we may find trees F_j with $S_j \subset F_j \subset E_j$. (Build up F_j by adding successive 'branches', to join each point of S_j by an arc in E_j to the part of F_j already constructed.) Then $F_j \to E$ in the Hausdorff metric.

Fix $\delta > 0$. Using Lemma 3.17 we may decompose each tree F_j so that $F_j = \bigcup_{i=1}^k F_{ji}$, where $|F_{ji}| \leq \delta$ for all i, j, and where $\sum_{i=1}^k |F_{ji}| \leq \mathcal{H}^1(F_j)$ for all j. (We may certainly use the same value of $k \leq 3\delta^{-1}c + 1$ for each j.) Applying the Blaschke selection theorem, Theorem 3.16, to $\{F_{ji}\}_j$ for each i in turn, we may assume, by taking subsequences and renumbering, that $\{F_{ji}\}_j$ converges in the Hausdorff metric to a non-empty compact set H_i for $1 \leq i \leq k$. Certainly, $|H_i| \leq \delta$ for all i, and also $E \subset \bigcup_1^k H_i$. Thus

$$\mathcal{H}^1_\delta(E) \leq \sum_i |H_i| = \lim_{j \to \infty} \sum_i |F_{ji}| \leq \varliminf_{j \to \infty} \mathcal{H}^1(F_j) \leq \varliminf_{j \to \infty} \mathcal{H}^1(E_j),$$

and the result follows on letting $\delta \to 0$. □

The following selection result is often useful in geometrical problems.

Corollary 3.19

Let $\mathscr{C}$ be an infinite collection of continua, all of $\mathcal{H}^1$-measure, at most, c and lying in a bounded portion of $\mathbb{R}^n$. Then there exists a sequence of distinct sets in $\mathscr{C}$ convergent in the Hausdorff metric to a continuum E with $\mathcal{H}^1(E) \leq c$.

Proof. We use Theorem 3.16 to obtain a sequence $\{E_j\}$ from $\mathscr{C}$ convergent to a non-empty compact set E. Theorem 3.18 ensures that E is a continuum with $\mathcal{H}^1(E) \leq c$. □

Besicovitch (1938) employs Theorem 3.18 and its corollary without full justification in his development of the theory of 1-sets, although such results had already been proved by Gołąb (1929). The above proof is given by Faber, Mycielski & Pedersen (1983) in connection with the problem of finding the shortest curve that meets all lines cutting a circle.

Some remarks on analogues of these theorems for higher-dimensional sets are included in Section 3.5.

3.3 Density and the characterization of regular 1-sets

We now restrict attention to subsets of the plane. In this section we derive an essential upper bound for the lower densities of an irregular 1-set and hence characterize regular and irregular 1-sets.

A 1-set is called a *Z-set* if its intersection with every rectifiable curve is of $\mathscr{H}^1$-measure zero.

It follows from Corollary 3.9 that an irregular set is a Z-set. We shall show that the converse is also true.

Lemma 3.20

The intersection of a Z-set with a continuum of finite $\mathscr{H}^1$-measure is of measure zero.

Proof. This is immediate from Theorem 3.14 and the definition of a Z-set.   □

The major part of this section is devoted to showing that a Z-set has upper density strictly less than 1 almost everywhere, and so is irregular. To this end we use Besicovitch's (1938, Section 15) idea of 'circle-pairs'. (A circle-pair is a figure formed by two discs of equal radii, each with its centre on the perimeter of the other.) The proof is complicated but rather remarkable. We follow the version of Besicovitch's proof given in the generalization of Morse & Randolph (1944, Section 9). We first require a topological result on the removal of the interiors of discs from continua. Recall that a collection of discs is *semidisjoint* if no member of the collection is included in any other.

Lemma 3.21

Let E be a continuum in $\mathbb{R}^2$. Suppose that $\{B_i\}_1^\infty$ is a countable semidisjoint collection of closed discs each contained in E and such that $|B_i| \geq d$ for only finitely many i for any $d > 0$. Then if Γ_i is the perimeter of B_i,

$$F = (E \setminus \bigcup_i B_i) \cup \bigcup_i \Gamma_i$$

is a continuum.

Proof. Note that $F = (E \setminus \bigcup_i \operatorname{int} B_i) \cup \bigcup_i \Gamma_i$, so to show F is closed it is enough to show that $\overline{\bigcup_i \Gamma_i} \subset F$. If $x \in \overline{\bigcup_i \Gamma_i} \setminus (E \setminus \bigcup_i \operatorname{int} B_i) \subset \bigcup_i \operatorname{int} B_i$, then

$x \in$ int B_k, say. Let $d > 0$ be the distance from x to the perimeter of B_k. If Γ_j intersects $B_{\frac{1}{4}d}(x)$, then not only does B_j meet $B_{\frac{1}{4}d}(x)$ but it also meets $\mathbb{R}^2 \backslash B_k$, since $B_j \not\subset B_k$. Thus $|\Gamma_j| \geq \frac{1}{2}d$, so only finitely many circles Γ_j can cut $B_{\frac{1}{4}d}(x)$. Since $x \in \bigcup_i \Gamma_i$ we conclude that for some $k, x \in \overline{\bigcup_{i=1}^k \Gamma_i} = \bigcup_{i=1}^k \Gamma_i \subset F$, as required.

To prove that F is connected, suppose that $F = F_1 \cup F_2$, where F_1 and F_2 are disjoint closed sets. Let

$$E_1 = F_1 \cup \bigcup_{\{i:\Gamma_i \subset F_1\}} B_i, \quad E_2 = F_2 \cup \bigcup_{\{i:\Gamma_i \subset F_2\}} B_i.$$

Every Γ_i is contained in either E_1 or E_2 so that $E_1 \cup E_2 = E$. Since the discs $\{B_i\}$ are semidisjoint and $F \cap B_i \subset \bigcup_i \Gamma_i$ for each i, it is easy to check that $E_1 \cap E_2 = \varnothing$. The set E_1 is closed, for if $x \in \overline{\bigcup_{\{i:\Gamma_i \subset F_1\}} B_i}$, then x is either the limit of a sequence of points in $\bigcup_1^k B_i$ for some k, or else the limit point of a sequence of discs with boundaries in the closed set F_1 and radii tending to 0. In either case $x \in F_1$. Similarly, E_2 is closed, so $E = E_1 \cup E_2$ is a disjoint decomposition of E into closed sets. Since E is connected either E_1 or E_2 is empty, and thus either F_1 or F_2 is empty. □

The substance of the proof of the upper bound for the lower densities of a Z-set is contained in the next lemma. We let $R(x, y)$ denote the *common region* of the circle-pair with centres x and y, so that $R(x, y) =$ int $(B_{|x-y|}(x) \cap B_{|x-y|}(y))$. Roughly speaking, we show that if the common regions of the circle-pairs with centres in a 1-set E contain very much of E, then it is possible to join up the various components of E by curves which lie to an appreciable extent in E.

Lemma 3.22
Let E be a 1-set in $\mathbb{R}^2$ and suppose that $\alpha > 0$. Let E_0 be a compact subset of E with $\mathscr{H}^1(E_0) > 0$, such that $\mathscr{H}^1(E \cap R(x_1, x_2)) \geq \alpha |x_1 - x_2|$ if $x_1, x_2 \in E_0$. Then there exists a continuum H such that $0 < \mathscr{H}^1(H \cap E) \leq \mathscr{H}^1(H) < \infty$.

Proof. By the usual process of uniformization we may, using Corollary 2.5, find $\rho_1 > 0$ and a compact set $F \subset E_0$ with $\mathscr{H}^1(F) > 0$ such that

$$\mathscr{H}^1(B_r(x) \cap E) \leq 2 \cdot 2r \quad \text{if } x \in F \quad \text{and } 0 < r \leq \rho_1. \tag{3.11}$$

Further, by Corollaries 2.4 and 2.5, there exists a point $y \in F$ and a number ρ with $0 < \rho \leq \frac{1}{10}\rho_1$ such that

$$\mathscr{H}^1((E\backslash F) \cap B_r(y)) < 2r \cdot 10^{-3}\alpha \quad (0 < r \leq 3\rho) \tag{3.12}$$

and

$$\mathscr{H}^1(F \cap B_\rho(y)) > \frac{1}{4} \cdot 2\rho = \frac{1}{2}\rho. \tag{3.13}$$

By reducing ρ if necessary, we may also assume that the perimiter Γ of the disc $B_\rho(y)$ contains some point of F.

Let $\mathscr{C}$ be the family of closed discs

$$\mathscr{C} = \{B_r(x) : x \in F \cap B_\rho(y), \quad 0 < r < 2\rho$$
$$\text{and } \mathscr{H}^1((E\backslash F) \cap B_r(x)) \geq \alpha r\}. \tag{3.14}$$

By Lemma 1.9 we may find a null, finite or countable disjoint collection of closed discs $\{B_i\}$ in $\mathscr{C}$ such that $\bigcup_{B \in \mathscr{C}} B \subset \bigcup_i B_i'$, where B_i' is the closed disc concentric with B_i and of five times the radius. Further, we may take the discs $\{B_i'\}$ to be semidisjoint. Write Γ_i for the perimeter of B_i'.

Letting

$$G = (F \cap B_\rho(y)) \cup \Gamma \cup \bigcup_i B_i',$$

we define

$$H = (G\backslash \bigcup_i B_i') \cup \bigcup_i \Gamma_i, \tag{3.15}$$

see Figure 3.2. The remainder of the proof, which is divided into six stages, checks that H is a continuum with the stated properties.

(a) *G is closed*: since $F \cap B_\rho(y)$ and Γ are closed, it is enough to show that $\bigcup_i B_i' \subset G$. But if $z \in \bigcup_i B_i'$, then either $z \in \bigcup_{i=1}^k B_i' \subset G$ for some $k < \infty$, or else z is the limit of points from a subsequence of the discs $\{B_i'\}_i$. On summing areas, $\sum_i |B_i'|^2 < \infty$, so $|B_i'| \to 0$ as $i \to \infty$, thus, z is the limit point of the centres of these discs which all lie in the closed set $F \cap B_\rho(y)$.

(b) *G is connected*: we suppose that $G = G_1 \cup G_2$, where G_1 and G_2 are non-empty disjoint closed sets, and derive a contradiction. Any connected

Fig. 3.2

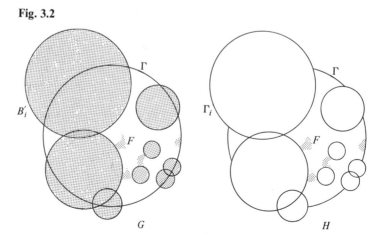

G H

subset of G must be contained in either G_1 or G_2, so assume that $\Gamma \subset G_1$. Each B_i' has centre in $B_\rho(y)$, so if B_i' contains points on or outside Γ it must meet Γ and therefore be contained in G_1. Thus $G_2 \subset \text{int } B_\rho(y)$.

Let G_1' be the set obtained by adjoining $\mathbb{R}^2 \backslash B_\rho(y)$ to G_1, so that G_1' is closed, connected and disjoint from G_2. Since both G_1' and G_2 contain points of F, let $x_1 \in F \cap G_1'$ and $x_2 \in F \cap G_2$ be points that minimize the distance $r = |x_1 - x_2|$. This infimum is attained and is positive. As $\Gamma \subset G_1'$ contains a point of F, $0 < r < 2\rho$. The common region $R(x_1, x_2)$ is disjoint from F, otherwise r could be further reduced, so from the hypotheses of the lemma

$$0 < \alpha r \leq \mathcal{H}^1(E \cap R(x_1, x_2)) = \mathcal{H}^1((E \backslash F) \cap R(x_1, x_2))$$
$$\leq \mathcal{H}^1((E \backslash F) \cap B_r(x_2)).$$

Since $x_2 \in F \cap B_\rho(y)$ and $r < 2\rho$, $B_r(x_2)$ is a member of $\mathscr{C}$ and so $B_r(x_2) \subset \bigcup_i B_i' \subset G$. Thus $B_r(x_2) = (B_r(x_2) \cap G_1') \cup (B_r(x_2) \cap G_2)$ is a decomposition of $B_r(x_2)$ into disjoint closed sets with x_1 in the first and x_2 in the second. Clearly, this is absurd, since $B_r(x_2)$ is connected.

(c) H is a continuum: by (a) and (b) G is a continuum, so by Lemma 3.21 H is also a continuum.

(d) $\sum_i |B_i'| \leq \frac{1}{8}\rho$: using (3.14), the disjointedness of the B_i, and (3.12), we see that

$$\sum_i |B_i'| = 5 \sum_i |B_i| \leq \frac{10}{\alpha} \sum_i \mathcal{H}^1((E \backslash F) \cap B_i)$$

$$\leq \frac{10}{\alpha} \mathcal{H}^1((E \backslash F) \cap B_{3\rho}(y)) \leq \frac{10}{\alpha} \cdot 6\rho 10^{-3} \alpha \leq \tfrac{1}{8}\rho.$$

(e) $\mathcal{H}^1(H) < \infty$: since $H \subset E \cup \Gamma \cup \bigcup_i \Gamma_i$, we have, using (d),

$$\mathcal{H}^1(H) \leq \mathcal{H}^1(E) + 2\pi\rho + \pi \sum_i |B_i'| < \infty.$$

(f) $\mathcal{H}^1(H \cap E) > 0$: from the definition of H

$$\mathcal{H}^1(H \cap E) \geq \mathcal{H}^1(G \cap E) - \mathcal{H}^1(E \cap \bigcup_i B_i')$$

$$\geq \mathcal{H}^1(F \cap B_\rho(y)) - \sum_i \mathcal{H}^1(E \cap B_i')$$

$$\geq \mathcal{H}^1(F \cap B_\rho(y)) - 2 \sum_i |B_i'|$$

$$\geq \tfrac{1}{2}\rho - \tfrac{1}{4}\rho = \tfrac{1}{4}\rho > 0,$$

using (3.11) (note that $|B_i'| < 20 \ \rho \leq 2\rho_1$), with (3.13) and (d) providing the final estimates. □

The proof of the theorem on the lower densities of Z-sets is now an easy consequence of the geometry of circle-pairs. We exploit the fact that a circle-pair with centres at points of high lower density and normal convex density must contain a subset of positive measure in its common region.

Theorem 3.23

Let E be a Z-set in $\mathbb{R}^2$. Then $\underline{D}^1(E, x) \leq \frac{3}{4}$ at almost all $x \in E$.

Proof. Suppose that for some $\alpha > 0$ the set $E_1 = \{x : \underline{D}^1(E, x) > \frac{3}{4} + \alpha\}$ has positive measure. Then, using Theorem 2.3, we may, in the usual way, find a compact 1-set $E_2 \subset E_1$ and $\rho > 0$ such that

$$\mathscr{H}^1(E \cap B_r(x)) > (\tfrac{3}{4} + \alpha)2r \quad (x \in E_2, 0 < r \leq \rho) \tag{3.16}$$

and

$$\mathscr{H}^1(E \cap U) < (1 + \alpha)|U| \quad (x \in E_2 \cap U, \quad 0 < |U| \leq 3\rho). \tag{3.17}$$

Let E_0 be a compact subset of E_2 with $0 < \mathscr{H}^1(E_0) < \infty$ and $|E_0| \leq \rho$. If $x_1, x_2 \in E_0$, then $r = |x_1 - x_2| \leq \rho$, so noting that any circle intersects E in a set of measure 0,

$$\begin{aligned}
\mathscr{H}^1(E \cap R(x_1, x_2)) &= \mathscr{H}^1(E \cap B_r(x_1)) + \mathscr{H}^1(E \cap B_r(x_2)) \\
&\quad - \mathscr{H}^1(E \cap (B_r(x_1) \cup B_r(x_2))) \\
&\geq 2(\tfrac{3}{4} + \alpha)2r - (1 + \alpha)3r = \alpha r = \alpha|x_1 - x_2|,
\end{aligned}$$

using (3.16) and (3.17) with $U = B_r(x_1) \cap B_r(x_2)$. Hence applying Lemma 3.22 to $E_0 \subset E$, we deduce that there exists a continuum H with $0 < \mathscr{H}^1(H \cap E) \leq \mathscr{H}^1(H) < \infty$. Since E is a Z-set this is impossible by Lemma 3.20. We conclude that $\mathscr{H}^1(E_1) = 0$ if $\alpha > 0$, as required. $\qquad \square$

Corollary 3.24

Let E be an irregular 1-set in $\mathbb{R}^2$. Then $\underline{D}^1(E, x) \leq \frac{3}{4}$ at almost all $x \in E$.

Proof. By Corollary 3.9 no subset of E of positive measure is contained in any rectifiable curve. Thus E is a Z-set and the result follows immediately from Theorem 3.23. $\qquad \square$

We have shown that, as far as lower density is concerned, regular and irregular 1-sets are of a very different nature, regular sets having lower density 1, and irregular sets having lower density of, at most, $\frac{3}{4}$ almost everywhere. It is *not* possible to construct a 1-set with lower density strictly between $\frac{3}{4}$ and 1 on a set of positive measure.

The exact essential upper bound of $\underline{D}^1(E, x)$ is still unknown. Whilst $\frac{3}{4}$ is the best value obtained to date, it is generally believed that $\frac{1}{2}$ is the correct answer; certainly sets may be constructed with $\underline{D}^1(E, x) = \frac{1}{2}$ almost everywhere. It is amusing to note that Besicovitch's (1928a) first estimate of the bound was $1 - 10^{-2576}$, before he reduced this to $\frac{3}{4}$ in his subsequent (1938) paper.

Besicovitch (1938) also developed his circle-pair method to show that the density of an irregular set fails to exist almost everywhere, see also Besicovitch & Walker (1931), Gillis (1934b) and Morse & Randolph (1944).

We can now characterize regular and irregular 1-sets.

Theorem 3.25
An irregular 1-set is a Z-set. A regular 1-set consists of a Y-set, together with a set of $\mathcal{H}^1$-measure zero.

Proof. If E is irregular, then E intersects every rectifiable curve Γ in a set of measure zero, for otherwise $E \cap \Gamma$ would be a Y-set and so would be regular.

If E is regular, then $\underline{D}^1(E, x) = 1$ almost everywhere, so Theorem 3.23 with Corollary 2.6 implies that any measurable subset of E of positive measure intersects some rectifiable curve in a set of positive measure. We use this fact to define a sequence of rectifiable curves $\{\Gamma_j\}$. Choose Γ_1 so that

$$\mathcal{H}^1(\Gamma_1 \cap E) \geq \tfrac{1}{2} \sup \{\mathcal{H}^1(\Gamma \cap E) : \Gamma \text{ is rectifiable}\}.$$

If $\Gamma_1, \ldots, \Gamma_k$ have been selected and $E_k = E \backslash \bigcup_1^k \Gamma_j$ is of positive measure, let Γ_{k+1} be a rectifiable curve with

$$\mathcal{H}^1(\Gamma_{k+1} \cap E_k) \geq \tfrac{1}{2} \sup \{\mathcal{H}^1(\Gamma \cap E_k) : \Gamma \text{ is rectifiable}\} > 0. \qquad (3.18)$$

The process terminates only if for some k the curves $\{\Gamma_j\}_1^k$ cover almost all of E, in which case the conclusion is clear. Otherwise,

$$\infty > \mathcal{H}^1(E) \geq \sum_k \mathcal{H}^1(\Gamma_{k+1} \cap E_k),$$

so that $\mathcal{H}^1(\Gamma_{k+1} \cap E_k) \to 0$. If $\mathcal{H}^1(E \backslash \bigcup_1^\infty \Gamma_j) > 0$, then we may find a rectifiable curve Γ with $\mathcal{H}^1(\Gamma \cap (E \backslash \bigcup_1^\infty \Gamma_j)) = d > 0$. But $\mathcal{H}^1(\Gamma_{k+1} \cap E_k) < \tfrac{1}{2}d$ for some k, and Γ would have been selected in preference to Γ_{k+1}, according to (3.18). Hence $\mathcal{H}^1(E \backslash \bigcup_1^\infty \Gamma_j) = 0$, so E consists of the Y-set $\bigcup_j (E \cap \Gamma_j)$ and a set of measure zero. $\square$

An irregular 1-set, as a Z-set, intersects every rectifiable curve in a set of measure 0. However, Besicovitch (1928a, Section 43) shows that an irregular set may be enclosed in a countable union of non-rectifiable curves.

Finally in this section we indicate the 'broken' nature of irregular sets. Recall that a set is *totally disconnected* if no two of its points lie in the same connected component. Thus, given any pair of points in the set, there is a decomposition into two disjoint closed subsets, each containing one of the points.

Corollary 3.26
An irregular 1-set E is totally disconnected.

Proof. If x and y both lie in the same closed connected subset F of E, then $|x - y| \leq \mathcal{H}^1(F) < \infty$ by Lemma 3.4. But E is a Z-set and so $\mathcal{H}^1(F) = \mathcal{H}^1(E \cap F) = 0$ by Lemma 3.20. Thus in fact $x = y$. $\qquad\square$

3.4 Tangency properties

As with the density properties, the tangency properties of regular sets and irregular sets are very different. Recalling the definition of a tangent (3.4), the result for regular sets is already clear:

Theorem 3.27

A regular 1-set E in $\mathbb{R}^n$ has a tangent at almost all of its points.

Proof. By Theorem 3.25, E is made up of a Y-set and a set of measure zero, and by Corollary 3.10 any Y-set has a tangent at almost all of its points.

$\qquad\qquad\qquad\qquad\qquad\qquad\qquad\qquad\qquad\qquad\qquad\qquad\qquad\square$

To investigate the tangency properties of irregular sets in the plane we use an ingenious argument of Besicovitch (1938, Section 10). We first require a geometrical lemma.

Lemma 3.28

Let θ be a unit vector in $\mathbb{R}^2$ perpendicular to a line L. Let P be a parallelogram with sides making angles ϕ to directions $\pm\theta$ and let y and z be opposite vertices of P, as in Figure 3.3. Then $|y - z| \leq d/\sin \phi$, where d is the length of projection of P onto L.

Fig. 3.3

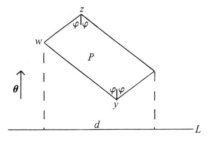

Proof. Let w be a third vertex of P. Then $|y - z| \leq |y - w| + |w - z|$ and $d = (|y - w| + |w - z|) \sin \phi$, leading to the result. $\qquad\square$

The next theorem is essentially Besicovitch's tangency result for irregular sets. The idea of the proof is to reduce an irregular set of points to a subset of a rectifiable curve which necessarily has zero measure.

Theorem 3.29

Let E be an irregular 1-set in $\mathbb{R}^2$. Then, given θ and $0 < \phi < \pi/2$,

$$\bar{D}^1(E, x, \theta, \phi) + \bar{D}^1(E, x, -\theta, \phi) \geq \tfrac{1}{6}\sin\phi \qquad (3.19)$$

for almost all $x \in E$.

Proof. Take $\rho, \delta_+, \delta_- > 0$, and let $F_0 = F_0(\delta_+, \delta_-, \rho)$ be the set of x in E for which both

$$\mathcal{H}^1(E \cap S_r(x, \theta, \phi)) \leq 2r\delta_+$$

and
$$\qquad (3.20)$$

$$\mathcal{H}^1(E \cap S_r(x, -\theta, \phi)) \leq 2r\delta_-$$

for all $r \leq \rho$. As usual, F_0 is measurable; we shall show that if F_0 has positive measure, then $\delta_+ + \delta_-$ cannot be too small.

If $\mathcal{H}^1(F_0) > 0$, then by regularity of $\mathcal{H}^1$ we may find a compact subset F_1 of F_0 with positive measure. Further, from Theorem 2.3, or directly from the definition of $\mathcal{H}^1$, we may, given $\eta > 0$, find a closed convex set U with $0 < |U| \leq \rho$ and

$$\mathcal{H}^1(F) > (1 - \eta)|U|, \qquad (3.21)$$

where $F = F_1 \cap U$. From now on we work inside U.

As F is closed, we may choose y_1 and z_1 to be (among) the most distant pair of points in F which have their connecting segment at an angle of not more than ϕ with θ, so that

$$r_1 = |y_1 - z_1| = \sup\{|y - z| : z \in F \cap S(y, \theta, \phi) \quad \text{and}$$
$$y \in F \cap S(z, -\theta, \phi)\},$$

where $S(y, \theta, \phi)$ denotes the infinite sector. The maximality of r_1 ensures that

$$F \cap S(y_1, \theta, \phi) = F \cap S_{r_1}(y_1, \theta, \phi)$$

and

$$F \cap S(z_1, -\theta, \phi) = F \cap S_{r_1}(z_1, -\theta, \phi).$$

Since $r_1 \leq \rho$ and $F \subset E$ we conclude from (3.20) that

$$\mathcal{H}^1(F \cap S(y_1, \theta, \phi)) \leq 2r_1\delta_+$$

and
$$\qquad (3.22)$$

$$\mathcal{H}^1(F \cap S(z_1, -\theta, \phi)) \leq 2r_1\delta_-.$$

Let P_1 be the closed parallelogram

$$P_1 = S(y_1, \theta, \phi) \cap S(z_1, -\theta, \phi)$$

Fig. 3.4

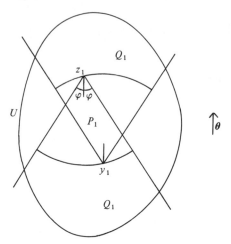

and let Q_1 be the open region

$$Q_1 = \text{int} \{S(y_1, \theta, \phi) \cup S(z_1, -\theta, \phi)\},$$

see Figure 3.4. From (3.22)

$$\mathcal{H}^1(F \cap Q_1) \leq 2r_1(\delta_+ + \delta_-).$$

The set $U \backslash Q_1$ is compact and has, at most, two components, so provided it contains a pair of points of F in the same component with joining segment at an angle of at most ϕ to θ, we may in the same way obtain a parallelogram P_2 outside Q_1, and an open region Q_2. Thus for $j = 1, 2, \ldots$ we find points y_j and z_j of F lying in the same component of $U \backslash \bigcup_{i=1}^{j-1} Q_i$, with $r_j = |y_j - z_j|$ as large as possible, and with

$$\mathcal{H}^1(F \cap Q_j) \leq 2r_j(\delta_+ + \delta_-), \tag{3.23}$$

where

$$Q_j = \text{int} \{S(y_j, \theta, \phi) \cup S(z_j, -\theta, \phi)\}$$

and

$$P_j = S(y_j, \theta, \phi) \cap S(z_j, -\theta, \phi).$$

This process either continues indefinitely or until no suitable pair of points is left.

Let L be a line perpendicular to θ. If $j > i$, then P_j is disjoint from Q_i, so the projections of the parallelograms $\{P_j\}$ onto L are segments with disjoint interiors. An easy check using Lemma 3.28 sees that

$$\sum_j r_j = \sum_j |y_j - z_j| \leq \frac{3|U|}{\sin \phi}; \tag{3.24}$$

the '3' allows for the possibility of up to two of the P_j overlapping either side of U.

Let y and z be distinct points of $F\backslash\bigcup_i Q_i$. If y and z lie on opposite sides of some Q_j, then $[y, z]$ makes an angle of at least ϕ with θ. If not, and if $[y, z]$ makes an angle of less than ϕ with θ, then y and z would have been candidates for y_j and z_j at each stage, so the process of selection could not have terminated and $|y - z| \leq |y_j - z_j|$ for all j. This is impossible as (3.24) implies $|y_j - z_j| \to 0$. We conclude that the angle made with L by the segment joining any pair of points of $F\backslash\bigcup_i Q_i$ is, at most, $\frac{1}{2}\pi - \phi$. By Lemma 3.11 $F\backslash\bigcup_i Q_i$ is a subset of a rectifiable curve. Thus as $F\backslash\bigcup_i Q_i$ is a subset of E, which is a Z-set by Lemma 3.25,

$$\mathcal{H}^1(F\backslash\bigcup_i Q_i) = 0.$$

Hence, using (3.23), (3.24) and (3.21),

$$\mathcal{H}^1(F) \leq \sum_j \mathcal{H}^1(F \cap Q_j) \leq 2(\delta_+ + \delta_-)\sum_j r_j$$

$$\leq \frac{6(\delta_+ + \delta_-)|U|}{\sin\phi} \leq \frac{6(\delta_+ + \delta_-)}{(1 - \eta)\sin\phi}\mathcal{H}^1(F).$$

Given any $\eta > 0$ there is some F with $\mathcal{H}^1(F) > 0$ for which this holds, so

$$(\delta_+ + \delta_-) \geq \tfrac{1}{6}\sin\phi.$$

Thus if $\delta_+ + \delta_- < \tfrac{1}{6}\sin\phi$ the set $F_0(\delta_+, \delta_-, \rho)$ defined by (3.20) must have measure zero, and the proof is complete. $\square$

Corollary 3.30
Let E be an irregular 1-set in $\mathbb{R}^2$. Then, for almost all $x \in E$,

$$\bar{D}^1(E, x, \theta, \phi) + \bar{D}^1(E, x, -\theta, \phi) \geq \tfrac{1}{6}\sin\phi > \tfrac{1}{10}\phi \tag{3.25}$$

for all θ and all $0 < \phi < \tfrac{1}{2}\pi$.

Proof. By Theorem 3.29, inequality (3.19) holds for some countable dense subset of $\{(\theta, \phi)\}$ for almost all x. Thus, by approximation, (3.19) holds for *all* θ and ϕ for almost all x. $\square$

The coefficient on the right-hand side of (3.19) or (3.25) can certainly be improved. In fact Besicovitch's proof, which uses slightly better estimates than those of Lemma 3.28 and (3.24), yields $\tfrac{1}{3}\sin\phi$, and indeed $\tfrac{1}{2}\sin\phi$ in case $\phi \geq \tfrac{2}{3}\pi$. Besicovitch conjectures that the best possible coefficient may be as large as $2^{-\frac{1}{2}}$ (an example in Besicovitch (1982a, Section 11) shows it cannot be larger than this). However, any positive value leads to the main tangency result.

Corollary 3.31
At almost all points of a plane irregular 1-set no tangent exists.

Proof. This is immediate from Corollary 3.30 and the definition (3.4) of a tangent.   □

As we have seen, regular 1-sets may be obtained as subsets of rectifiable curves. Examples of irregular sets are perhaps less obvious, so we now give a simple construction of an irregular 1-set.

Theorem 3.32
There exist irregular 1-sets in $\mathbb{R}^2$.

Proof. Let f be the real function on $[0, 1)$ defined by

$$f(.x_1 x_2 \ldots) = .y_1 y_2 \ldots,$$

where $.x_1 x_2 \ldots$ and $.y_1 y_2 \ldots$ are written to base 4 (with the convention that the x_i are not ultimately all equal to 3) so that $y_i = 5 - x_i$ (mod 4). (See Figure 3.5.) Let

$$E = \{(x, f(x)) : 0 \leq x < 1\}.$$

Fig. 3.5

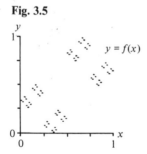

E is a Borel set and, since the projection of E onto the x-axis has unit length, it follows that $1 \leq \mathcal{H}^1(E)$. On the other hand, E may be covered by 4^k squares of side 4^{-k} for every k, so that $\mathcal{H}^1(E) \leq \sqrt{2}$. Thus E is a 1-set. It is easy to see that any rectifiable curve (which has a tangent almost everywhere) intersects E in length zero. Hence E is a Z-set and is irregular. Alternatively, irregularity follows from the fact that E projects onto sets of Lebesgue measure zero in the two directions $\pm 45°$ to the x-axis; see Corollary 6.14.   □

3.5 Sets in higher dimensions

Generalization of the theory to integral-dimensional subsets of $\mathbb{R}^n$ $(n \geq 3)$ was not completed until some 47 years after the publication of Besicovitch's first fundamental paper (1928*a*). This was because of the considerable technical difficulties that had to be overcome; for example, the theorem that a continuum of finite $\mathscr{H}^s$-measure is a union of rectifiable curves has no analogue if s is an integer greater than 1.

An s-set in $\mathbb{R}^n$ is called *countably rectifiable* if it is of the form

$$\bigcup_{j=1}^{\infty} f_j(E_j) \cup G,$$

where $\mathscr{H}^s(G) = 0$ and where each f_j is a Lipschitz function from a bounded subset E_j of $\mathbb{R}^s$ to $\mathbb{R}^n$. (A Lipschitz function f requires that $|f(t) - f(u)| \leq c|t - u|$ for some constant c.) This is a direct analogue of Besicovitch's 'Y-set, together with a set of measure zero'.

An s-set E in $\mathbb{R}^n$ has a *tangent* at x if $\bar{D}^s(E, x) > 0$ and there is an s-dimensional plane Π through x such that for all $\phi > 0$,

$$\lim_{r \to 0} r^{-s} \mathscr{H}^s(E \cap (B_r(x) \backslash S(x, \Pi, \phi))) = 0, \tag{3.26}$$

where $S(x, \Pi, \phi)$ denotes the set of $y \in \mathbb{R}^n$ with $[y, x]$ making an angle of at most ϕ with Π. If $\underline{D}^s(E, x) > 0$ and 'lim' in (3.26) is replaced by '$\underline{\lim}$', then we say E has a *weak tangent* at x.

Theorem 3.33
Let E be an s-set in $\mathbb{R}^n$ where s is an integer. Then the following statements are equivalent:

(a) *E is regular.*
(b) *E is countably rectifiable.*
(c) *E has a tangent at almost all of its points.*

As we have seen, Besicovitch (1928a, 1938) showed this for 1-sets in $\mathbb{R}^2$. Federer (1947) demonstrated the equivalence of (b) and (c) in an enormously complicated paper, and it follows from this and the definition of the tangent that (b) implies (a). (Federer's concept of restrictedness is equivalent to the existence of a tangent.) Moore (1950) completed the proofs for 1-sets in $\mathbb{R}^n$, Marstrand (1961) for 2-sets in $\mathbb{R}^3$, and finally Mattila (1975a) for s-sets in $\mathbb{R}^n$. Results of this nature have also been obtained for more general measures μ. Here we must define regularity simply as the existence of the density $\lim_{r \to \infty} r^{-s}\mu(B_r(x))$ μ-almost everywhere, without reference to the actual value of the limit. The papers of Federer and Moore mentioned above consider such generalizations, as do Morse & Randolph (1944), for subsets of $\mathbb{R}^2$ with $s = 1$, and Marstrand (1964) in the general case. In particular, Marstrand shows that for general measures, regularity implies that weak tangents (defined in the obvious way) exist almost everywhere; it is natural to conjecture that the word 'weak' can be omitted.

Many of these proofs are intimately connected with higher-dimensional analogues of the projection properties discussed in Chapter 6.

There are considerable difficulties in extending the theory of curves to s-dimensional surfaces in $\mathbb{R}^n$ if s is an integer and $1 < s < n$. The theory is far more complicated than for curves; indeed great care is required with the precise meaning that is attached to terms such as 'surface' and 'area'.

Surfaces in $\mathbb{R}^3$ have been studied for many years, and for a long time the accepted definition of surface area was the Lebesgue–Fréchet definition as the supremum of the areas of inscribed polyhedra, by analogy with the definition of the length of a curve. It was Besicovitch (1945b) who demonstrated that this definition was hopelessly inadequate, not even possessing the basic additive properties of a measure. He proposed that the most satisfactory definition of area was the Hausdorff $\mathscr{H}^2$-measure, and set about tackling a variety of problems on surface area from this standpoint.

One question that arises is how to generalize Gołąb's semicontinuity theorem, Theorem 3.18, to surfaces. Clearly, there is no direct analogue, since it is possible to approximate to any surface in the Hausdorff metric by a sequence of rectifiable curves which must surely have zero area. Besicovitch (1949) showed that $\mathscr{H}^2(E) \leq \varliminf_{j \to \infty} \mathscr{H}^2(E_j)$ if $\{E_j\}$ is a sequence of parametric surfaces convergent to the parametric surface E, provided that the limiting surface E satisfies reasonable smoothness conditions. A more recent generalization of Gołąb's theorem to higher-dimensional sets is due to Vitushkin (1966), see also Ivanov (1975). We define the *variations* of a compact subset E of $\mathbb{R}^n$ as follows. Let $V^s(E)$ be the number of connected components of E, so that if $V^s(E) = 1$ then E is a continuum. For $1 \leq s \leq n$ let $V^s(E) = c \int V^0(E \cap \Pi) \mathrm{d}\Pi$, where Π is an $(n-s)$-dimensional plane and integration is with respect to the natural invariant measure on the space of all such planes (as occurs in integral geometry). The normalizing constant c depends only on n and s and is chosen so that if E is an s-dimensional cube then $V^s(E)$ is its volume. Then, for a sufficiently smooth, or polyhedral surface, V^s is simply s-dimensional volume. Originally these variations were used in the study of convex sets, but they may be applied with advantage to general compact sets. Vitushkin's semicontinuity theorem states that if $\{E_j\}$ is a sequence of compact sets in $\mathbb{R}^n$ convergent to E in the Hausdorff metric, then $V^s(E) \leq \varliminf_{j \to \infty} V^s(E_j)$ provided that the numbers $\{V^i(E_j)\}$ are uniformly bounded for $0 \leq i < s$. If $s = 1$ this reduces to Gołąb's theorem.

A problem of special interest is to find 'minimal surfaces', that is, surfaces of minimum area that span a closed curve in $\mathbb{R}^3$. A careful definition of 'surface' is required and then the problem lends itself to treatment using

Hausdorff measures. The 'Plateau problem' of whether a surface of minimum area always exists was finally solved in the affirmative by Reifenberg (1960). (This paper also lists earlier references.) A survey of some of these topics is given by Besicovitch (1950).

If E is a bounded Lebesgue-measurable subset of $\mathbb{R}^n$ with $\mathscr{L}^n(E) > 0$, the *density boundary* ∂E of E consists of those points in $\mathbb{R}^n$ at which the Lebesgue density of E either fails to exist or else takes a value other than 0 or 1. The Lebesgue density theorem, Theorem 1.13, implies that $\mathscr{L}^n(\partial E) = 0$. However, in the other direction, it is possible to show that $\mathscr{H}^{n-1}(\partial E) > 0$ a non-trivial result if E has a highly irregular boundary.

A variety of other surface measures have been devised. As well as those already mentioned, those in common use include the integral-geometric measures and the de Gorgi perimeter; see Federer (1969) where such measures are defined and examined in detail.

Exercises on Chapter 3

3.1   Let $\psi : [a, b] \to \mathbb{R}^n$ be a (not necessarily continuous) function of bounded variation, that is, with
$$\mathrm{var}\,[a, b] = \sup \sum_{i=1}^{m} |\psi(t_i) - \psi(t_{i-1})| < \infty,$$
where the supremum is over all dissections $a = t_0 < t_1 < \cdots < t_m = b$ of $[a, b]$. Show that $\psi[a, b]$ is contained in a (possibly self-intersecting) rectifiable curve. (Hint: Use $\mathrm{var}\,[a, t]$ to define a continuous function from $[0, \mathrm{var}\,[a, b]]$ to $\mathbb{R}^n$.) Deduce that $\mathscr{H}^1(\psi[a, b]) < \infty$.

3.2   Prove that any arcwise connected set is connected. Exhibit a continuum in $\mathbb{R}^2$ that is not arcwise connected. (Compare Lemma 3.12.)

3.3   Verify that the Hausdorff metric is a metric.

3.4   Give an example of a sequence of continua $\{E_j\}$ in $\mathbb{R}^2$ convergent to E in the Hausdorff metric, such that $\mathscr{H}^1(E_j) \to \infty$ but with $\mathscr{H}^1(E) < \infty$.

3.5   Let E be a continuum with $\mathscr{H}^1(E) < \infty$. Show that there exists a tree $F \subset E$ with $\mathscr{H}^1(E \backslash F)$ as small as desired. Deduce that E is contained in a (possibly self-intersecting) curve of length at most $2\mathscr{H}^1(E)$.

3.6   Let E be an irregular 1-set. Given $\varepsilon, c > 0$, show that there exists $\delta_0 > 0$ such that $\mathscr{H}^1(E \cap [F]_\delta) < \varepsilon$ if $0 < \delta < \delta_0$ for any continuum F with $\mathscr{H}^1(F) < c$, where $[F]_\delta$ is the δ-parallel body of F.

3.7   Prove that there exists a compact set $F \subset \mathbb{R}^2$ of Hausdorff dimension 2 such that every 1-set contained in F is irregular.

3.8   Construct a totally disconnected *regular* 1-set in $\mathbb{R}^2$. (Compare Corollary 3.26.)

4
Structure of sets of non-integral dimension

4.1 Introduction

This chapter examines local properties of s-sets in $\mathbb{R}^n$ for non-integral s. The fundamental result is that any such set is irregular, that is, has lower circular or spherical density strictly less than 1 at almost all of its points. Indeed, the stronger result that its density fails to exist at almost all of its points has also been established. As before, we also examine the existence of suitably defined tangents, and show that the set of points at which such tangents exist must have measure zero.

For the case of subsets of the plane, the work is entirely due to Marstrand (1954a, 1955), the former paper providing a very complete account. As with sets of integral dimension, higher-dimensional analogues present formidable difficulties; the natural generalizations were eventually proved by Marstrand (1964).

4.2 s-sets with $0 < s < 1$

First we consider s-sets in $\mathbb{R}^n$ for s strictly less than 1. In this case the basic properties, including non-existence of the density almost everywhere, are relatively easy to obtain.

The following topological observation about such sets is sometimes useful.

Lemma 4.1
An s-set E in $\mathbb{R}^n$ with $0 < s < 1$ is totally disconnected.

Proof. Let x and y be distinct points in the same connected component of E. Define a mapping $f : \mathbb{R}^n \to [0, \infty)$ by $f(z) = |z - x|$. Since f does not increase distances it follows from Lemma 1.8 that $\mathscr{H}^s(f(E)) \leq \mathscr{H}^s(E) < \infty$. As $s < 1$ it follows that $f(E)$ is a subset of $\mathbb{R}$ of Lebesgue measure zero and, in particular, has dense complement. Choosing a number r with $r \notin f(E)$ and $0 < r < f(y)$ we have

$$E = \{z \in E : |z - x| < r\} \cup \{z \in E : |z - x| > r\}.$$

This is an open decomposition of E with x in one part and y in the other, contradicting the assumption that x and y belong to the same connected component of E. □

The following proof of the density property is a slightly shortened version of that of Marstrand (1954a).

Theorem 4.2
If E is an s-set with $0 < s < 1$ *the density* $D^s(E, x)$ *fails to exist at almost every point of E.*

Proof. If the conclusion is false, E has a measurable subset of positive measure where the density exists and is at least $2^{-s} > \frac{1}{2}$, by Corollary 2.5. Choosing ρ small enough we may find an s-set $F \subset E$ such that if $x \in F$, then $D^s(E, x)$ exists and

$$\mathscr{H}^s(E \cap B_r(x)) > \tfrac{1}{2}(2r)^s \tag{4.1}$$

for all $r \leq \rho$. By regularity of $\mathscr{H}^s$ we may further assume that F is closed. Let y be an accumulation point of F and let η be a number with $0 < \eta < 1$. If $A_{r,\eta}$ denotes the annular region $B_{r(1+\eta)}(y) \backslash B_{r(1-\eta)}(y)$, then

$$
(2r)^{-s}\mathscr{H}^s(E \cap A_{r,\eta}) = (2r)^{-s}\mathscr{H}^s(E \cap B_{r(1+\eta)}(y)) - (2r)^{-s}\mathscr{H}^s(E \cap B_{r(1-\eta)}(y))
$$
$$
\to D^s(E, y)((1+\eta)^s - (1-\eta)^s) \tag{4.2}
$$

as $r \to 0$. On the other hand, for arbitrarily small values of r we may find $x \in F$ with $|x - y| = r$. Then

$$B_{\frac{1}{2}r\eta}(x) \subset A_{r,\eta},$$

so (4.1) gives

$$\tfrac{1}{2}r^s\eta^s < \mathscr{H}^s(E \cap B_{\frac{1}{2}r\eta}(x)) \leq \mathscr{H}^s(E \cap A_{r,\eta})$$

Fig. 4.1

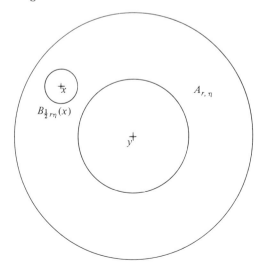

for arbitrarily small values of r (see Figure 4.1). Using (4.2) we conclude that

$$2^{-(s+1)}\eta^s \le D^s(E, y)((1 + \eta)^s - (1 - \eta)^s) = D^s(E, y)(2s\eta + 0(\eta^2))$$

as $\eta \to 0$. This is impossible if $s < 1$ and the theorem follows from this contradiction.   □

Corollary 4.3
Any s-set with $0 < s < 1$ is irregular.

The following result of Marstrand (1954a) on lower angular densities has a proof in some ways akin to that of Theorem 4.2.

Theorem 4.4
Let θ be a unit vector and let $\phi < \frac{1}{2}\pi$. Then if E is an s-set with $0 < s < 1$ we have $\underline{D}^s(E, x, \theta, \phi) = 0$ at almost all $x \in E$.

Proof. Suppose to the contrary. Then in the usual way we may find ρ_0, $\alpha > 0$ and a closed s-set $F \subset E$ such that if $x \in F$ and $r \le \rho_0$,

$$\mathscr{H}^s(E \cap S_r(x, \theta, \phi)) > \alpha r^s. \tag{4.3}$$

By Corollaries 2.4 and 2.5 we may certainly choose $y \in F$ such that $D^s(E\backslash F, y) = 0$ and $\bar{D}^s(E, y) = c2^{-s}$, where $0 < c < \infty$. Consequently, given $\varepsilon > 0$, there exists $\rho_1 \le \rho_0$ such that if $r \le \rho_1$,

$$\mathscr{H}^s((E\backslash F) \cap B_r(y)) < \varepsilon r^s \tag{4.4}$$

and

$$\mathscr{H}^s(E \cap B_r(y)) < (c + \varepsilon) r^s. \tag{4.5}$$

In addition, choose $\rho \le \frac{1}{2}\rho_1$ such that

$$(c - \varepsilon)\rho^s < \mathscr{H}^s(E \cap B_\rho(y)). \tag{4.6}$$

Fig. 4.2

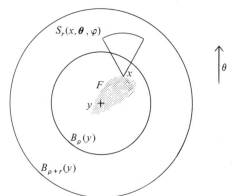

Let x be a point of $F \cap B_\rho(y)$ that maximizes the scalar product $x \cdot \theta$, see Figure 4.2. Then

$$(F \cap B_\rho(y)) \cup (F \cap S_r(x, \theta, \phi)) \subset F \cap B_{\rho+r}(y)$$

with the union disjoint except for the point x. Thus

$$\mathcal{H}^s(F \cap B_\rho(y)) + \mathcal{H}^s(F \cap S_r(x, \theta, \phi)) \leq \mathcal{H}^s(F \cap B_{\rho+r}(y)),$$

so

$$\mathcal{H}^s(E \cap B_\rho(y)) + \mathcal{H}^s(E \cap S_r(x, \theta, \phi))$$
$$\leq \mathcal{H}^s(E \cap B_{\rho+r}(y)) + 2\mathcal{H}^s((E \backslash F) \cap B_{\rho+r}(y)).$$

If $r \leq \rho$, then $\rho + r \leq \rho_1$ so we may use (4.3)–(4.6) to see that

$$(c - \varepsilon)\rho^s + \alpha r^s < (c + \varepsilon)(\rho + r)^s + 2\varepsilon(\rho + r)^s.$$

On writing $\gamma = r/\rho$ this becomes

$$c - \varepsilon + \alpha \gamma^s < (c + 3\varepsilon)(1 + \gamma)^s,$$

true for $0 < \gamma < 1$. This holds for all $\varepsilon > 0$, so

$$c + \alpha \gamma^s \leq c(1 + \gamma)^s \leq c + cs\gamma$$

for $0 < \gamma < 1$. This contradicts the positivity of α. $\qquad \square$

This result remains true for $\phi = \frac{1}{2}\pi$ provided that the densities are calculated taking $S_r(x, \theta, \frac{1}{2}\pi)$ as the *open* semicircle. Davies has observed that this follows as a direct n-dimensional analogue of the result on one-sided densities in $\mathbb{R}$, elegantly proved by Besicovitch (1968) in his penultimate paper.

The question of tangents to s-sets where $0 < s < 1$ is not of particular interest as the sets are so sparse as to make any idea of approximation by line segments rather meaningless.

In one sense it is possible for a tangent to exist at all points. Let Γ be a smooth curve in $\mathbb{R}^n$ defined by a smooth bijection $\psi : \mathbb{R} \to \mathbb{R}^n$ and let E be an s-set in $\mathbb{R}$. It is easy to see that $\psi(E)$ is an s-set in $\mathbb{R}^n$ and that for almost all $x \in \psi(E)$ the tangent to Γ at x is a tangent to $\psi(E)$ at x in the sense of (3.4).

4.3 *s*-sets with $s > 1$

If $s > 1$ the density and tangency properties of s-sets are much more complicated, and we restrict detailed exposition to subsets of the plane.

We deal with tangency questions first. One approach is to repeat Besicovitch's proof of Theorem 3.29 to deduce that the sum of the upper angular densities of a plane s-set $(1 < s < 2)$ in any opposite pair of angles is strictly positive almost everywhere. Besicovitch's proof depends on the intersection of the set under consideration with any rectifiable curve having measure zero; this is certainly true if $s > 1$ by the remark following Lemma 3.2. Thus it follows from the definition of a tangent (3.4) that a plane s-set with $1 < s < 2$ fails to have a tangent almost everywhere. However, it is easy to prove a stronger result, and when it comes to questions of circular density, we require this more delicate information on tangential behaviour.

We follow Marstrand (1954a) in defining a weak tangent. An s-set E has a *weak tangent* in direction θ at x if $D^s(x, E) > 0$ and if for every $\phi > 0$,

$$\lim_{r \to 0} r^{-s} \mathcal{H}^s(E \cap (B_r(x) \backslash S_r(x, \theta, \phi) \backslash S_r(x, -\theta, \phi))) = 0. \tag{4.7}$$

(Recall $S_r(x, \theta, \phi)$ is the sector consisting of those points in $B_r(x)$ which make an angle of, at most, ϕ at x with the half-line from x in direction θ.)

Unlike the tangent defined in (3.4) it is possible in principle for a set to have many weak tangents at a point. However, this does not usually happen.

Lemma 4.5

Let E be an s-set in $\mathbb{R}^2$ with $1 < s < 2$. Then for almost all $x \in E$,

$$\bar{D}^s(E, x, \theta, \phi) \leq 4.10^s \phi^{s-1}$$

for all θ and all $\phi \leq \frac{1}{2}\pi$.

Proof. Fix $\rho > 0$ and let

$$F = \{x \in E : \mathcal{H}^s(E \cap B_r(x)) < 2^{s+1} r^s \text{ for all } r \leq \rho\}. \tag{4.8}$$

Choose $x \in F$ and any θ and ϕ with $0 < \phi \leq \frac{1}{2}\pi$. For $i = 1, 2, \ldots$ let A_i be the intersection of annulus and sector

$$A_i = S_{ir\phi}(x, \theta, \phi) \backslash S_{(i-1)r\phi}(x, \theta, \phi),$$

so that $S_r(x, \theta, \phi) \subset \bigcup_1^m A_i \cup \{x\}$ for an integer m less than $2/\phi$. The diameter of each set A_i is at most $10r\phi < \rho$ if $r < \rho/20$, so applying (4.8) to each A_i that contains points of F and summing,

$$\mathcal{H}^s(F \cap S_r(x, \theta, \phi)) \leq 2\phi^{-1} 2^{s+1} (10r\phi)^s$$

or

$$(2r)^{-s} \mathcal{H}^s(F \cap S_r(x, \theta, \phi)) \leq 4 \cdot 10^s \phi^{s-1},$$

if $r < \rho/20$. Thus $\bar{D}^s(F, x, \theta, \phi) \leq 4 \cdot 10^s \phi^{s-1}$ at almost all $x \in F$, so, since by Corollary 2.6 $\bar{D}^s(E \backslash F, x) = 0$ at almost all $x \in F$, the conclusion of the lemma holds at almost all $x \in F$. By Corollary 2.5 almost all points of E lie in such an F for some $\rho > 0$, hence the result.   □

Corollary 4.6

If E is an s-set in $\mathbb{R}^2$ with $1 < s < 2$, then at almost all points of E no weak tangent exists.

Proof. By Lemma 4.5,

$$\lim_{r \to 0} (2r)^{-s} \mathcal{H}^s(E \cap (B_r(x) \backslash S_r(x, \theta, \phi) \backslash S_r(x, -\theta, \phi)))$$

$$\geq \underline{D}^s(E, x) - \bar{D}^s(E, x, \theta, \phi) - \bar{D}^s(E, x, -\theta, \phi)$$

$$\geq \underline{D}^s(E, x) - 8 \cdot 10^s \phi^{s-1}$$

for all θ and $\phi < \frac{1}{2}\pi$ for almost all $x \in E$. At such points either $\underline{D}^s(E, x) = 0$ or (4.7) fails to hold for ϕ sufficiently small, so no weak tangent exists. $\qquad\square$

To show that an s-set is irregular if $1 < s < 2$, we first consider angular densities rather than circular densities.

Theorem 4.7
Let E be an s-set in $\mathbb{R}^2$ with $1 < s < 2$. Then if $\phi < \frac{1}{2}\pi$ the lower angular density $\underline{D}^s(E, x, \theta, \phi)$ is zero for some θ for almost all $x \in E$.

Proof. Fix positive numbers α and ρ and let

$$F_0 = \{x : \mathcal{H}^s(E \cap S_r(x, \theta, \phi)) > \alpha r^s \text{ for all}$$

$$r \leq \rho \text{ and for all } \theta\}. \tag{4.9}$$

We show that $\mathcal{H}^s(F_0) = 0$. Suppose not, then by Corollary 2.5 we may find $\rho_1 \leq \rho$ and a set $F \subset F_0$ of positive measure such that if $x \in F$ and $r \leq \rho_1$, then

$$\mathcal{H}^s(E \cap B_r(x)) < 2^{s+1} r^s. \tag{4.10}$$

By regularity of $\mathcal{H}^s$ we may also assume F to be closed.

Let y be a point of F at which the circular density of $E \backslash F$ is zero; by Corollary 2.6 almost any point of F will suffice. Thus, given $\varepsilon > 0$, we may find $\rho_2 \leq \rho_1$ such that

$$\mathcal{H}^s((E \backslash F) \cap B_r(y)) < \varepsilon r^s \tag{4.11}$$

if $r \leq \rho_2$. We now work inside the disc $B_{\rho_2}(y)$. First we show that there are points in $B_{\frac{1}{2}\rho_2}(y)$ relatively remote from the set F. Suppose that for some $\gamma \leq \frac{1}{2}\rho_2$ all points of $B_{\frac{1}{2}\rho_2}(y)$ are within distance γ of F. Then if $x \in B_{\frac{1}{2}\rho_2}(y)$, there is a point z of F inside $B_\gamma(x)$. By (4.9) for any θ,

$$\alpha \gamma^s < \mathcal{H}^s(E \cap S_\gamma(z, \theta, \phi)) \leq \mathcal{H}^s(E \cap B_\gamma(z))$$

$$\leq \mathcal{H}^s(E \cap B_{2\gamma}(x)). \tag{4.12}$$

If $\gamma < \frac{1}{4}\rho_2$, then $B_{\rho_2}(y)$ contains $(\rho_2/\gamma)^2/16$ disjoint discs with centres in $B_{\frac{1}{2}\rho_2}(y)$ and radii 2γ. Thus, summing (4.12) over these discs,

$$(\rho_2/\gamma)^2 \alpha \gamma^s / 16 < \mathcal{H}^s(E \cap B_{\rho_2}(y)) < 2^{s+1} \rho_2^s,$$

by (4.10), so $\gamma > c\rho_2$, where c depends only on α and s. Thus if $\gamma \leq c\rho_2$ there is a disc of radius γ contained in $B_{\rho_2}(y)$ and containing no points of F. Hence

Fig. 4.3

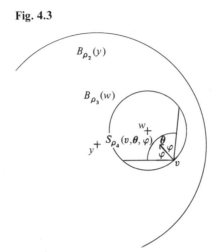

we may find a disc $B_{\rho_3}(w) \subset B_{\rho_2}(y)$ with no points of F in its interior but with its boundary containing a point v of F_0, with

$$\rho_2 \geq \rho_3 \geq c\rho_2. \tag{4.13}$$

Let θ be the inward normal direction to $B_{\rho_3}(w)$ at v, and let ρ_4 be half the length of the chords of $B_{\rho_3}(w)$ through v that make angles ϕ with θ (see Figure 4.3). By elementary geometry,

$$\rho_4 = \rho_3 \cos \phi. \tag{4.14}$$

As the sector $S_{\rho_4}(v, \theta, \phi)$ lies in $B_{\rho_3}(w)$ it contains no points of F other than v. Thus

$$\begin{aligned}
\mathscr{H}^s(E \cap S_{\rho_4}(v, \theta, \phi)) &= \mathscr{H}^s((E\backslash F) \cap S_{\rho_4}(v, \theta, \phi)) \\
&\leq \mathscr{H}^s((E\backslash F) \cap B_{\rho_2}(y)) \\
&< \varepsilon\rho_2^s \\
&< \varepsilon c_1 \rho_4^s,
\end{aligned} \tag{4.15}$$

by (4.11), (4.13) and (4.14), with c_1 dependent only on ϕ, α and s. Hence for any $\varepsilon > 0$ we may find $v \in F_0$ and ρ_4 with $0 < \rho_4 < \rho$ and θ for which (4.15) holds. This contradicts the definition (4.9) of F_0, so we conclude that $\mathscr{H}^s(F_0) = 0$. The observation that we may do this for any $\alpha, \rho > 0$ completes the proof.   □

Corollary 4.8

Let E be an s-set in $\mathbb{R}^2$ with $1 < s < 2$. Then at almost every point of E the lower angular density $\underline{D}^s(E, x, \theta, \frac{1}{2}\pi)$ is zero for some θ.

Proof. Take a sequence $\{\phi_i\}$ increasing to $\frac{1}{2}\pi$. By Theorem 4.7 we may, for almost all $x \in E$, find a sequence of directions $\{\theta_i\}$ such that

$D^s(E, x, \theta_i, \phi_i) = 0$ for each i. Extracting a convergent subsequent, we may assume that $\theta_i \to \theta$. It follows that $\underline{D}^s(E, x, \theta, \phi) = 0$ for any $\phi < \frac{1}{2}\pi$. Thus, provided x is also not one of the exceptional points for Lemma 4.5, we see that $\underline{D}^s(E, x, \theta, \frac{1}{2}\pi) = 0$ $\square$

We next examine lower angular densities of *regular s-sets* in $\mathbb{R}^2$ for $1 < s < 2$, as a step towards proving that such sets do not exist. The following ingenious argument of Marstrand (1954a) shows, roughly speaking, that if such a set is sparse to one side of a line through one of its points, then it must also be sparse on the other side.

Lemma 4.9
Let E be an s-set in $\mathbb{R}^2$, where $1 < s < 2$. Let x be a regular point of E at which the upper convex density equals 1, and suppose that $\underline{D}^s(E, x, -\theta, \frac{1}{2}\pi) = 0$ for some θ. Then E has a weak tangent at x perpendicular to θ.

Proof.
Since $D^s(E, x) = 1$ and $\bar{D}_c^s(E, x) = 1$ we may, given any $\eta > 0$, find arbitrarily small values of ρ such that

$$\mathcal{H}^s(E \cap B_r(x)) > 2^s r^s (1 - \eta) \quad \text{if } r \le \rho, \tag{4.16}$$

$$\mathcal{H}^s(E \cap U) < (1 + \eta)|U|^s \quad \text{if } x \in U \text{ and } 0 < |U| \le 2\rho \tag{4.17}$$

and

$$\mathcal{H}^s(E \cap S_\rho(x, -\theta, \tfrac{1}{2}\pi)) < 2^s \eta \rho^s. \tag{4.18}$$

Take $0 < \phi < \frac{1}{2}\pi$, let L be the line through x perpendicular to θ, and let M and M' be the half-lines from x at angle ϕ to θ. For a fixed positive integer m, we construct inductively a sequence of $m + 1$ semicircles S_{r_i} of radius r_i, each with centre x and based on L, where $\rho = r_0 > r_1 > \ldots > r_m$. For each i the semicircle S_{r_i} will have ends y_i and y_i' on L and cut M and M' at z_i and z_i'. Suppose S_{r_i} has been constructed. Then $S_{r_{i+1}}$ is specified by taking y_{i+1} to be the point on $[x, y_i]$ such that its distance from y_i' equals the sum of its distances from y_i and z_i'. (A straightforward continuity argument shows

Fig. 4.4

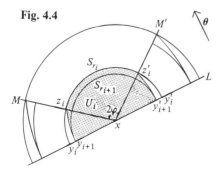

that such a point exists.) Draw an arc with centre y_{i+1} through z_i' to meet L at y_{i+1}'. By symmetry the arc with centre y_{i+1}' through z_i meets L at y_{i+1} (see Figure 4.4). We denote the (shaded) convex part of S_{r_i} cut off by these arcs by U_i. Then

$$|U_i| = 2r_{i+1}. \tag{4.19}$$

We estimate the measure of E contained between two consecutive sectors bounded by M and M'.

$$S_{r_i}(x, \theta, \phi) \backslash S_{r_{i+1}}(x, \theta, \phi) \subset U_i \backslash S_{r_{i+1}}(x, \theta, \tfrac{1}{2}\pi)$$
$$\subset U_i \cup S_\rho(x, -\theta, \tfrac{1}{2}\pi) \backslash B_{r_{i+1}}(x),$$

so

$$\mathcal{H}^s(E \cap S_{r_i}(x, \theta, \phi)) - \mathcal{H}^s(E \cap S_{r_{i+1}}(x, \theta, \phi))$$
$$\leq \mathcal{H}^s(E \cap U_i) + \mathcal{H}^s(E \cap S_\rho(x, -\theta, \tfrac{1}{2}\pi))$$
$$- \mathcal{H}^s(E \cap B_{r_{i+1}}(x))$$
$$< (1 + \eta)|U_i|^s + 2^s \eta \rho^s - 2^s r_{i+1}^s (1 - \eta)$$
$$\leq 2^{s+1} \eta r_{i+1}^s + 2^s \eta \rho^s$$
$$\leq 2^{s+2} \eta \rho^s,$$

using (4.16)–(4.19). Summing this estimate over all m sectors,

$$\mathcal{H}^s(E \cap S_\rho(x, \theta, \phi)) < 2^{s+2} \eta \rho^s m + \mathcal{H}^s(E \cap S_{r_m}(x, \theta, \phi))$$
$$< 2^{s+2} \eta \rho^s m + 2^s (1 + \eta) r_m^s,$$

by (4.17). By virtue of the construction, r_m/ρ depends only on ϕ and m, and this ratio tends to zero as m tends to infinity. Thus, given $\varepsilon > 0$, we may find m independent of η such that

$$\mathcal{H}^s(E \cap S_\rho(x, \theta, \phi)) < \rho^s(2^{s+2} \eta m + \varepsilon(1 + \eta)).$$

This holds for arbitrarily small values of ρ for any $\eta > 0$, so $\underline{D}^s(E, x, \theta, \phi) = 0$ if $\phi < \tfrac{1}{2}\pi$. Since $\underline{D}^s(E, x, -\theta, \tfrac{1}{2}\pi) = 0$, it follows from (4.7) that E has a weak tangent at x in a direction perpendicular to θ. □

Corollary 4.10
Let E be an s-set in $\mathbb{R}^2$ with $1 < s < 2$. Then E is irregular.

Proof. At almost all points of E the upper convex density equals 1, by Theorem 2.3, and $\underline{D}^s(E, x, \theta, \tfrac{1}{2}\pi) = 0$ for some θ, by Corollary 4.8. Hence by Lemma 4.9 the set E has a weak tangent at almost all of its regular points. This is inconsistent with Corollary 4.6 unless the set of regular points has measure zero, in which case E is an irregular s-set. □

Shortly after publishing the above result, Marstrand (1955) extended the method to prove the stronger fact that the density of an s-set in $\mathbb{R}^2 (1 < s < 2)$ fails to exist at almost all of its points.

For the sake of completeness we gather together the main results of this chapter.

Theorem 4.11
A plane s-set is irregular unless s is an integer.

Proof. This combines Corollary 4.3 and Corollary 4.10.   □

4.4 Sets in higher dimensions
The principal higher-dimensional result is similar to Theorem 4.11.

Theorem 4.12
An s-set in $\mathbb{R}^n$ is irregular unless s is an integer.

This has been proved for $0 < s < 1$ in Theorem 4.2, and the proof of Corollary 4.10 adapts to cover the case $n - 1 < s < n$. However, for intermediate values of s, substantial new ideas were required, and the theorem was eventually proved by Marstrand (1964). Indeed, Marstrand's work goes much farther than this, not only in that it applies to more general measures, but also in that it shows that the density fails to exist almost everywhere if s is non-integral. The proof involves a much deeper study of weak tangential properties of regular sets.

Exercises on Chapter 4

4.1   Adapt the proof of Theorem 4.2 to show that if E is an s-set with $0 < s < 1$, then $\underline{D}^s(E, x) \leq (1 + 2^{s/(s-1)})^{s-1}$ for almost all x.

4.2   Let E be an s-set ($0 < s < 1$). Deduce from Theorem 4.4 that for any unit vector θ the lower hemispherical density $\underline{D}^s(E, x, \theta, \frac{1}{2}\pi) = 0$ at almost all $x \in E$ provided that the density is calculated using the open semicircle or hemisphere. Give an example to show that this is false for the closed semicircle.

4.3   Prove the following special case of Theorem 4.12: let E be an s-set in $\mathbb{R}^3$ lying in a smooth (infinitely differentiable, say) surface. Then E is irregular unless s is an integer.

5
Comparable net measures

5.1 Construction of net measures

Comparable net measures are an extremely useful tool in the study of Hausdorff measures. Net measures behave much more conveniently than Hausdorff measures but can nevertheless be constructed to be equivalent for many purposes. First used by Besicovitch (1952) in his demonstration that closed sets of infinite $\mathscr{H}^s$-measure contain subsets of positive but finite measure, they were later employed by Marstrand (1954b) in work on the Hausdorff measure of Cartesian products of sets. In this chapter we are particularly concerned with these two applications.

We restrict attention here to a basic form of net measure on Euclidean space. For a wider ranging discussion see Section 2.7 of Rogers (1970) describing the work of Davies and Rogers.

Net measures are constructed in a similar manner to Hausdorff measures but using a restricted class $\mathscr{N}$ of covering sets in the definition rather than the class of all sets. The class $\mathscr{N}$ is chosen to be a 'net' of sets, with the essential property that if $U_1, U_2 \in \mathscr{N}$, then either $U_1 \cap U_2 = \varnothing$ or $U_1 \subset U_2$ or $U_2 \subset U_1$. Moreover, we assume that each set of $\mathscr{N}$ is contained in finitely many others. In particular, given any collection of sets in $\mathscr{N}$, it is possible, by removing those sets contained in any others, to find a disjoint subcollection with the same union. As we shall see, we may construct net measures 'comparable' to Hausdorff measures, that is, with the ratio of the measures bounded above and below.

To construct the s-dimensional net measure $\mathscr{M}^s$ on $\mathbb{R}^n$ let $\mathscr{N}$ be the collection of all n-dimensional half-open binary cubes, specifically, sets of the form

$$[2^{-k}m_1, 2^{-k}(m_1 + 1)) \times [2^{-k}m_2, 2^{-k}(m_2 + 1))$$
$$\times \ldots \times [2^{-k}m_n, 2^{-k}(m_n + 1)),$$

where k is a non-negative integer and $m_1, \ldots, m_n$ are integers. (If $n = 1$ or 2 the net $\mathscr{N}$ consists of half-open binary intervals or squares.) If $E \subset \mathbb{R}^n$ and $\delta > 0$ define

$$\mathscr{M}^s_\delta(E) = \inf \sum_{i=1}^{\infty} |S_i|^s, \tag{5.1}$$

where the infimum is over all countable δ-covers of E by sets $\{S_i\}$ of $\mathscr{N}$. By the net property it is sufficient to consider covers of E by disjoint collections of sets of $\mathscr{N}$. Then $\mathscr{M}_\delta^s(E)$ is an outer measure on $\mathbb{R}^n$ that is finite on bounded subsets of $\mathbb{R}^n$. Letting

$$\mathscr{M}^s(E) = \lim_{\delta \to 0} \mathscr{M}_\delta^s(E) = \sup_{\delta > 0} \mathscr{M}_\delta^s(E)$$

we obtain a metric outer measure $\mathscr{M}^s$ on $\mathbb{R}^n$. By Theorem 1.5 the Borel sets are $\mathscr{M}^s$-measurable (as indeed are the Souslin sets). A proof similar to that of Theorem 1.6 shows that $\mathscr{M}^s$ is a regular outer measure.

We demonstrate that $\mathscr{M}^s$ and $\mathscr{H}^s$ are indeed comparable measures; this is essentially the work of Besicovitch (1952).

Theorem 5.1

There exist constants b_n dependent only on the dimension n such that for every $E \subset \mathbb{R}^n$,

$$\mathscr{H}_\delta^s(E) \le \mathscr{M}_\delta^s(E) \le b_n \mathscr{H}_\delta^s(E) \tag{5.2}$$

if $0 < \delta < 1$, and

$$\mathscr{H}^s(E) \le \mathscr{M}^s(E) \le b_n \mathscr{H}^s(E). \tag{5.3}$$

Proof. It is immediate from the definitions of the outer measures that $\mathscr{H}_\delta^s(E) \le \mathscr{M}_\delta^s(E)$. (For Hausdorff measures the infimum is taken over a larger class of covering sets.)

If U is any set with $0 < |U| \le \delta$, let k be the integer such that

$$2^{-k-1} \le |U| < 2^{-k}, \tag{5.4}$$

and let S be a binary cube of side 2^{-k} that intersects U. Then U is contained in the collection of 3^n binary cubes of side 2^{-k} and diameter $2^{-k}n^{1/2}$ consisting of S and its immediate neighbours. Subdividing each of these cubes into 2^{n^2} smaller cubes, U is contained in $b_n = 3^n 2^{n^2}$ binary cubes of diameter

$$2^{-k}n^{1/2}2^{-n} \le 2^{1-n}n^{1/2}|U| \le |U| \le \delta, \tag{5.5}$$

using (5.4). Now let $\{U_i\}$ be a δ-cover of E by arbitrary sets. For each i we have $U_i \subset \bigcup_{j=1}^{b_n} S_{ij}$, where $\{S_{ij}\}_{j=1}^{b_n}$ is a collection of b_n cubes of diameter, at most, $|U_i| \le \delta$.

Thus $E \subset \bigcup_i \bigcup_j S_{ij}$ and

$$\sum_{i=1}^{\infty} \sum_{j=1}^{b_n} |S_{ij}|^s \le b_n \sum_{i=1}^{\infty} |U_i|^s,$$

so the right-hand inequality of (5.2) follows from the definitions of $\mathscr{M}_\delta^s$ and $\mathscr{H}_\delta^s$.

Inequality (5.3) follows on letting $\delta \to 0$. □

We conclude this section by proving two technical lemmas on monotonic sequences of sets which we will require later in the chapter.

Lemma 5.2
Let $\{E_j\}$ be a decreasing sequence of compact subsets of $\mathbb{R}^n$. Then, for any $\delta > 0$.

$$2^s \mathcal{H}^s_\delta(\lim_{j \to \infty} E_j) \geq \lim_{j \to \infty} \mathcal{H}^s_{2\delta}(E_j). \tag{5.6}$$

Proof. Let $\{U_i\}$ be any δ-cover of $\lim E_j$. For each i let V_i be an open set containing U_i with $|V_i| \leq 2|U_i|$ and let V denote the open set $\bigcup_i V_i$. We claim that $E_j \subset V$ for some integer j. Otherwise, $\{E_j \backslash V\}$ is a decreasing sequence of non-empty compact sets, which, by an elementary consequence of compactness, has a non-empty limit set $(\lim E_j) \backslash V$. Then

$$2^s \sum |U_i|^s \geq \sum |V_i|^s \geq \mathcal{H}^s_{2\delta}(E_j) \geq \lim_{j \to \infty} \mathcal{H}^s_{2\delta}(E_j),$$

and (5.6) follows by considering all such δ-covers of $\lim E_j$.   □

If $\{E_j\}$ is any increasing sequence of subsets of $\mathbb{R}^n$, then $\lim_{j \to \infty} \mathcal{M}^s(E_j) = \mathcal{M}^s(\lim_{j \to \infty} E_j)$ by Lemma 1.3. However, this conclusion also holds for the outer measures $\mathcal{M}^s_\delta$. This is a particular case of the 'increasing sets lemma' which is of considerable importance in the theory of Hausdorff measures and is discussed in detail in Rogers (1970, Section 2.6), Davies (1970) and the references contained therein. We require the following special case.

Lemma 5.3
Let $\{E_j\}$ be an increasing sequence of subsets of $\mathbb{R}^n$ and suppose that each E_j is a finite union of binary cubes. Then

$$\mathcal{M}^s_\delta(\lim_{j \to \infty} E_j) = \lim_{j \to \infty} \mathcal{M}^s_\delta(E_j).$$

Proof. Write $E = \lim_{j \to \infty} E_j = \bigcup_j E_j$. Since $\mathcal{M}^s_\delta(E_j) \leq \mathcal{M}^s_\delta(E)$ for all j, it is enough to prove that $\mathcal{M}^s_\delta(E) \leq \lim_{j \to \infty} \mathcal{M}^s_\delta(E_j)$ on the assumption that the right-hand side is finite. As E_j is a finite union of binary cubes, the infimum in the definition of $\mathcal{M}^s_\delta(E_j)$ is attained by some finite disjoint collection of binary cubes. Suppose for each j that $\mathcal{S}_j$ is such a finite δ-cover of E_j, so that

$$\sum_{C \in \mathcal{S}_j} |C|^s = \mathcal{M}^s_\delta(E_j) \tag{5.7}$$

and assume that $\mathscr{S}_j$ is one of the numerically smallest collections of cubes with this property. If $S \in \mathscr{S}_j$, then S must contain a point of E_j. This point also lies in E_{j+1} and thus in some $T \in \mathscr{S}_{j+1}$. By the net property either S is a subset of T or T is a subset of S. If T is a proper subset of S, then we may either replace S by the cubes of $\mathscr{S}_{j+1}$ that are contained in S to reduce $\sum_{C \in \mathscr{S}_j} |C|^s$, or else replace the cubes of $\mathscr{S}_{j+1}$ that are contained in S by the single cube S to reduce either $\sum_{C \in \mathscr{S}_{j+1}} |C|^s$ or the number of terms in this sum. We conclude that if $S \in \mathscr{S}_j$ there exists $T \in \mathscr{S}_{j+1}$ with $S \subset T$.

Let $\{C_1, C_2, \ldots\}$ be the set of cubes obtained from $\bigcup_{j=1}^{\infty} \mathscr{S}_j$ by excluding any cube contained in some other cube of the collection. Then $E_j \subset \bigcup_{i=1}^{\infty} C_i$ for each j so that $E \subset \bigcup_{i=1}^{\infty} C_i$ and

$$\mathscr{M}_\delta^s(E) \le \sum_{i=1}^{\infty} |C_i|^s. \tag{5.8}$$

But each cube C_i belongs to $\mathscr{S}_j$ for all sufficiently large j, from the conclusion of the previous paragraph. Thus, given k, we may find $j(k)$ such that the cubes $C_1, \ldots, C_k$ all lie in $\mathscr{S}_{j(k)}$. Using (5.8) and (5.7),

$$\mathscr{M}_\delta^s(E) \le \lim_{k \to \infty} \sum_{i=1}^{k} |C_i|^s \le \lim_{k \to \infty} \sum_{C \in \mathscr{S}_{j(k)}} |C|^s$$

$$= \lim_{k \to \infty} \mathscr{M}_\delta^s(E_{j(k)}) \le \lim_{j \to \infty} \mathscr{M}_\delta^s(E_j),$$

as required. ☐

5.2 Subsets of finite measure

One consequence of the results of this section is that there is an abundance of s-sets, that is, sets of positive *finite* $\mathscr{H}^s$-measure. In fact, given any topologically respectable set E with $\mathscr{H}^s(E) > 0$, we can find a compact subset F of E with $0 < \mathscr{H}^s(F) < \infty$. If E is not of σ-finite $\mathscr{H}^s$-measure this is far from trivial. (A *σ-finite* set is one that may be expressed as a countable union of sets of finite measure.) Such results were first obtained by Besicovitch (1952) for closed sets, and were immediately extended to Souslin sets by Davies (1952b).

Theorem 5.4

Let E be a closed subset of $\mathbb{R}^n$ with $\mathscr{H}^s(E) = \infty$.

(a) *Let c be a positive number. Then there is a compact subset F of E such that $\mathscr{H}^s(F) = c$.*

(b) *There is a compact subset F of E such that $\mathscr{H}^s(F) > 0$ and*

$$\mathscr{H}^s(B_r(x) \cap F) \le br^s \qquad (x \in \mathbb{R}^n, r \le 1)$$

for some constant b.

Proof. To keep the notation relatively simple we present the proof for

$n = 1$. For higher-dimensional cases the procedure is identical but with binary cubes replacing binary intervals. We may assume that E is bounded. As $\mathcal{M}^s(E) = \infty$ we may find an integer m such that

$$\mathcal{M}^s_{2^{-m}}(E) \geq 2^s c b_1,$$

where b_1 is the constant of Theorem 5.1.

We define inductively a decreasing sequence $\{E_k\}^\infty_m$ of closed subsets of E. Let $E_m = E$. For $k \geq m$ we define E_{k+1} by specifying its intersection with each binary interval of length 2^{-k}. Let I be such an interval. If $\mathcal{M}^s_{2^{-(k+1)}}(E_k \cap I) \leq 2^{-sk}$, simply take $E_{k+1} \cap I = E_k \cap I$. Then

$$\mathcal{M}^s_{2^{-(k+1)}}(E_{k+1} \cap I) = \mathcal{M}^s_{2^{-k}}(E_k \cap I) \leq 2^{-sk}, \tag{5.9}$$

since using I as a covering interval in calculating $\mathcal{M}^s_{2^{-k}}$ gives an estimate at least as large as taking intervals of smaller length. On the other hand, if $\mathcal{M}^s_{2^{-(k+1)}}(E_k \cap I) > 2^{-sk}$ choose $E_{k+1} \cap I$ to be a compact subset of $E_k \cap I$ with $\mathcal{M}^s_{2^{-(k+1)}}(E_{k+1} \cap I) = 2^{-sk}$. Such a subset exists as $\mathcal{M}^s_{2^{-(k+1)}}(E_{k+1} \cap I \cap (-\infty, u])$ is continuous in u. Since $\mathcal{M}^s_{2^{-k}}(E_k \cap I) = 2^{-sk}$, (5.9) again holds. We now sum (5.9) over all binary intervals I of length 2^{-k}. Any covering intervals for calculating $\mathcal{M}^s_{2^{-k}}$ and $\mathcal{M}^s_{2^{-(k+1)}}$ must be contained in some such I, so this gives

$$\mathcal{M}^s_{2^{-(k+1)}}(E_{k+1}) = \mathcal{M}^s_{2^{-k}}(E_k) \qquad (k \geq m).$$

Iterating,

$$\mathcal{M}^s_{2^{-k}}(E_k) = \mathcal{M}^s_{2^{-m}}(E_m) \qquad (k \geq m). \tag{5.10}$$

Let I be a binary interval of length 2^{-k}. If $m \leq k < r$, then $E_r \subset E_{k+1}$, so, using (5.9),

$$\mathcal{M}^s_{2^{-(k+1)}}(E_r \cap I) \leq \mathcal{M}^s_{2^{-(k+1)}}(E_{k+1} \cap I) \leq 2^{-sk}.$$

Thus, in calculating $\mathcal{M}^s_{2^{-k}}(E_r)$, any interval I of length 2^{-k} may be replaced by intervals of lengths, at most, $2^{-(k+1)}$ without increasing the infimum value. Hence

$$\mathcal{M}^s_{2^{-(k+1)}}(E_r) = \mathcal{M}^s_{2^{-k}}(E_r) \qquad (m \leq k < r),$$

so, iterating and incorporating (5.10),

$$\mathcal{M}^s_{2^{-m}}(E_r) = \mathcal{M}^s_{2^{-r}}(E_r) = \mathcal{M}^s_{2^{-m}}(E_m) \qquad (r \geq m). \tag{5.11}$$

Let F be the compact set $F = \bigcap^\infty_m E_k$. Then using Theorem 5.1 and (5.10),

$$\mathcal{H}^s(F) \leq \mathcal{M}^s(F) = \lim_{k \to \infty} \mathcal{M}^s_{2^{-k}}(F)$$

$$\leq \lim_{k \to \infty} \mathcal{M}^s_{2^{-k}}(E_k) = \mathcal{M}^s_{2^{-m}}(E_m) < \infty. \tag{5.12}$$

On the other hand, using Lemma 5.2, Theorem 5.1 and (5.11),

$$2^s \mathcal{H}^s(F) \geq 2^s \mathcal{H}^s_{2^{-(m+1)}}(F) \geq \lim_{k \to \infty} \mathcal{H}^s_{2^{-m}}(E_k)$$

$$\geq b_1^{-1} \lim_{k \to \infty} \mathcal{M}^s_{2^{-m}}(E_k)$$

$$= b_1^{-1} \mathcal{M}^s_{2^{-m}}(E_m) \geq 2^s c.$$

Combining these inequalities,

$$c \leq \mathcal{H}^s(F) < \infty.$$

If there is strict inequality on the left, a set of the form $F \cap [u, \infty)$ will have measure exactly c, using the continuity of the measure $\mathcal{H}^s$ from above and below (Theorem 1.1).

A slight variant of the above method proves (b). Let J be a closed binary interval of length 2^{-r}, where $r \geq m$. If we sum (5.9) over the binary subintervals I of length 2^{-k} that lie in J ($k \geq r$) and proceed as before, we get

$$\mathcal{M}^s_{2^{-k}}(E_k \cap J) = \mathcal{M}^s_{2^{-m}}(E_m \cap J) \qquad (k \geq r \geq m)$$

in place of (5.10), and thus

$$\mathcal{H}^s(F \cap J) \leq \mathcal{M}^s_{2^{-r}}(E_r \cap J) \leq \mathcal{M}^s_{2^{-r}}(J) \leq |J|^s$$

in place of (5.12). Any interval J_0 may be enclosed in, at most six consecutive binary intervals of lengths, at most, $|J_0|$, and this leads to the result. The full details of (b) are left to the reader. $\qquad \square$

In view of Theorem 5.4 an obvious question to ask is whether such results hold for more general classes of set. In his original paper, Besicovitch extended the theorem to $F_{\sigma\delta\sigma}$-sets. Davies (1952*b*) took the work to its natural conclusion by proving it for all Souslin sets, using the following intermediate result.

Theorem 5.5
Let E be any Souslin subset of $\mathbb{R}^n$ with $\mathcal{H}^s(E) = \infty$. Then E contains a closed subset of infinite $\mathcal{H}^s$-measure.

Proof. See Davies (1952*b*) or Rogers (1970). $\qquad \square$

Combining this with Theorem 5.4 we get:

Theorem 5.6
Let E be a Souslin subset of $\mathbb{R}^n$ with $\mathcal{H}^s(E) = \infty$.
(a) *For any positive constant c there is a compact subset F of E such that $\mathcal{H}^s(F) = c$.*

(b) *There is a compact subset F of E such that* $\mathscr{H}^s(F) > 0$ *and*

$$\mathscr{H}^s(B_r(x) \cap F) \leq br^s \qquad (x \in \mathbb{R}^n, r \leq 1)$$

for some constant b.

Finally in this section we mention a result of a more pathological nature, namely that if E is a compact set of non-σ-finite $\mathscr{H}^s$-measure, then E is so large that it may be decomposed into continuum-many subsets, each of non-σ-finite $\mathscr{H}^s$-measure. Davies (1968) and Rogers (1970, Section 2.8) provide further details of this and related ideas.

5.3 Cartesian products of sets

For notational convenience we restrict our attention throughout this section to subsets of the plane, though the work extends to higher dimensions without difficulty. Here (x, y) represents Cartesian coordinates in $\mathbb{R}^2$. If E is any subset of $\mathbb{R}^2$ we denote by E_x the section consisting of those points of E which have first coordinate equal to x.

If E is a Lebesgue-measurable subset of $\mathbb{R}^2$ an immediate consequence of Fubini's theorem is that E_x is $\mathscr{L}^1$-measurable for almost all x and

$$\mathscr{L}^2(E) = \int \mathscr{L}^1(E_x) \mathrm{d}\mathscr{L}^1(x),$$

where $\mathscr{L}^1$ and $\mathscr{L}^2$ are 1- and 2-dimensional Lebesgue measure. It is natural to ask if any such results hold with Hausdorff measures of fractional dimension replacing Lebesgue measures. For example, if A and B are subsets of the two coordinate axes, are $\mathscr{H}^{s+t}(A \times B)$ and $\mathscr{H}^s(A)\mathscr{H}^t(B)$ equal, at least to within constant multiple? In fact, as a later example shows, such results are far from true. A rudimentary attempt at a proof might involve covering A and B by intervals of length δ, inducing a covering of $A \times B$ by squares of side δ, and thus estimating the various Hausdorff measures. Unfortunately, interval lengths that result in an 'efficient' cover of A may give an 'inefficient' cover of B, making such estimates useless. (Of course, the product measure $\mathscr{H}^s \times \mathscr{H}^t$ need not be a Hausdorff measure.) It is, however, possible to obtain inequalities such as

$$\mathscr{H}^{s+t}(A \times B) \geq b\mathscr{H}^s(A)\mathscr{H}^t(B),$$

This was shown under certain conditions by Besicovitch & Moran (1945) and other early proofs based on density ideas, were given by Moran (1946, 1949), Freilich (1950) and Eggleston (1950a, 1953b); see also Eggleston (1950b). However, these proofs are rather technical and also impose restrictions on A and B. The following very general and elegant approach using comparable net measures is due to Marstrand (1954b).

The proof depends on the following combinatorial lemma.

Lemma 5.7

Let A be any subset of $\mathbb{R}$, let $\{I_i\}$ be a countable δ-cover of A by binary intervals, and let $\{a_i\}$ be a sequence of positive numbers. Suppose c is a constant such that

$$\sum_{\{i : x \in I_i\}} a_i > c \tag{5.13}$$

for all $x \in A$. Then

$$\sum_i a_i |I_i|^s \geq c \mathcal{M}^s_\delta(A). \tag{5.14}$$

Proof. First assume that the collections $\{I_i\}$ and $\{a_i\}$ are finite. By reducing the a_i slightly, we may take each a_i to be rational without losing (5.13), then by multiplying through by a denominator-clearing factor we may further assume that the a_i are integers. Thus taking a_i copies of the interval I_i for each i it is enough to assume $a_i = 1$ for all i.

Under these assumptions (5.13) implies that each x in A lies in at least $\lceil c \rceil$ of the intervals, where $\lceil c \rceil$ is the least integer greater than c. As $A \subset \bigcup_i I_i$ and as the $\{I_i\}$ are binary intervals having the net property, we may, by taking all those intervals not contained in any other intervals in the collection, choose a non-overlapping set of intervals containing A, say $\{I_i\}_{i \in \mathscr{S}_1}$. The remaining intervals still cover A at least $\lceil c \rceil - 1$ times over, so in the same way we may choose a non-overlapping subcollection of these, $\{I_i\}_{i \in \mathscr{S}_2}$, that also covers A. Proceeding in this manner we obtain $\lceil c \rceil$ collections of non-overlapping intervals, $\{I_i\}_{i \in \mathscr{S}_j}$ for $j = 1, 2, \ldots, \lceil c \rceil$, with each collection covering A, and with the sets of indices $\mathscr{S}_j$ pairwise disjoint. As $|I_i| \leq \delta$ for all i,

$$\sum_{i \in \mathscr{S}_j} |I_i|^s \geq \mathcal{M}^s_\delta(A),$$

so that, summing over all j, we obtain (5.14) in the case where $\{I_i\}$ is a finite collection of intervals.

Finally, if $\{I_i\}_1^\infty$ is an infinite set of binary intervals, set

$$A_k = \left\{ x \in \mathbb{R} : \sum_{\substack{i : x \in I_i \\ i \leq k}} a_i > c \right\}$$

for each k. From the finite case,

$$\sum_1^k a_i |I_i|^s \geq c \mathcal{M}^s_\delta(A_k).$$

But each A_k is a finite union of binary intervals, and the sequence $\{A_k\}$ is increasing with $A \subset \bigcup_k A_k = \lim_{k \to \infty} A_k$. By Lemma 5.3,

$$\sum_1^\infty a_i |I_i|^s \geq c \lim_{k \to \infty} \mathcal{M}^s_\delta(A_k) = c \mathcal{M}^s_\delta(\lim_{k \to \infty} A_k) \geq c \mathcal{M}^s_\delta(A),$$

completing the proof in the infinite case. $\qquad \square$

Theorem 5.8

Let E be a plane set and let A be any subset of the x-axis. Suppose that if $x \in A$, then $\mathscr{H}^t(E_x) > c$, for some constant c. Then

$$\mathscr{H}^{s+t}(E) \geq bc\,\mathscr{H}^s(A),$$

where b depends only on s and t.

Proof. In view of Theorem 5.1 it is enough to prove the result with '$\mathscr{H}$' replaced by '$\mathscr{M}$' throughout.

Given $\delta > 0$, let $\{S_i\}_1^\infty$ be a collection of binary squares forming a $2^{\frac{1}{2}}\delta$-cover of E. For each $x \in A$,

$$E_x \subset \bigcup_1^\infty (S_i)_x,$$

so

$$\mathscr{M}_\delta^t(E_x) \leq \sum_1^\infty |(S_i)_x|^t.$$

If $A_\delta = \{x \in A : \mathscr{M}_\delta^t(E_x) > c\}$, then for $x \in A_\delta$,

$$c < \sum_1^\infty |(S_i)_x|^t = 2^{-\frac{1}{2}t} \sum_{\{i\,:\,x \in \mathrm{proj}\,S_i\}} |S_i|^t,$$

where $\mathrm{proj}\,S_i$ is the binary interval, of length, at most, δ, obtained by projecting S_i onto the x-axis. Thus

$$\sum_1^\infty |S_i|^{s+t} = \sum_1^\infty |S_i|^t |S_i|^s = 2^{\frac{1}{2}s} \sum_1^\infty |S_i|^t |\mathrm{proj}\,S_i|^s$$

$$\geq 2^{\frac{1}{2}(s+t)} c\,\mathscr{M}_\delta^s(A_\delta),$$

taking $I_i = \mathrm{proj}\,S_i$ and $a_i = |S_i|^t$ in Lemma 5.7. But this is true for any $2^{\frac{1}{2}}\delta$-cover of E by binary squares $\{S_i\}$, so

$$bc\,\mathscr{M}_\delta^s(A_\delta) \leq \mathscr{M}_{2^{\frac{1}{2}}\delta}^{s+t}(E) \leq \mathscr{M}^{s+t}(E),$$

where $b = 2^{\frac{1}{2}(s+t)}$. Since A_δ increases to A as δ decreases to 0,

$$\mathscr{M}_\delta^s(A_\rho) \leq \mathscr{M}_\delta^s(A_\delta) \leq b^{-1}c^{-1}\mathscr{M}^{s+t}(E)$$

if $\delta \leq \rho$. Thus for $\rho > 0$,

$$\mathscr{M}^s(A_\rho) \leq b^{-1}c^{-1}\mathscr{M}^{s+t}(E)$$

and, by the continuity of the measure $\mathscr{M}^s$,

$$\mathscr{M}^s(A) \leq b^{-1}c^{-1}\mathscr{M}^{s+t}(E). \qquad \square$$

Now take E to be the Cartesian product $A \times B$.

Corollary 5.9

For any subsets A and B of $\mathbb{R}$,

$$\mathscr{H}^{s+t}(A \times B) \geq b\,\mathscr{H}^s(A)\mathscr{H}^t(B).$$

We may interpret this in terms of Hausdorff dimension:

Corollary 5.10
For any subsets A and B of $\mathbb{R}$,
$$\dim (A \times B) \geq \dim A + \dim B.$$

Notice that these results also hold if $t = 0$ when $\mathscr{H}^t(E_x)$ equals the number of points in E_x. The value of the constant b is discussed by Ernst & Freilich (1976).

In general, the inequalities in Theorem 5.8 and its corollaries may not be reversed, though Besicovitch & Moran (1945) and Taylor (1952) provide some sufficient conditions on A and B for equality in Corollary 5.10. The following demonstration that inequality may be strict is offered as a simple alternative to the example of Besicovitch & Moran (1945).

Theorem 5.11
There exist Borel subsets A and B of $\mathbb{R}$ of Hausdorff dimension 0 such that $\mathscr{H}^1(A \times B) > 0$.

Proof. Let $\{s_j\}$ be a sequence of numbers decreasing to 0 and let $0 = m_0 < m_1 < m_2 < \ldots$ be a sequence of integers increasing rapidly enough to ensure that

$$\left. \begin{aligned} (m_1 - m_0) + (m_3 - m_2) + \cdots + (m_{2j-1} - m_{2j-2}) \leq s_j m_{2j} \\ (m_2 - m_1) + (m_4 - m_3) + \cdots + (m_{2j} - m_{2j-1}) \leq s_j m_{2j+1}. \end{aligned} \right\} \quad (5.15)$$

Let A be the subset of $[0,1]$ consisting of those numbers with zero in the rth decimal place if $m_j + 1 \leq r \leq m_{j+1}$ and j is odd. Similarly, take B as the set of numbers which have zeros in the rth decimal place if $m_j + 1 \leq r \leq m_{j+1}$ and j is even. Taking the obvious covers of A by 10^k intervals of length $10^{-m_{2j}}$, where

$$k = (m_1 - m_0) + (m_3 - m_2) + \cdots + (m_{2j-1} - m_{2j-2}),$$

it follows from (5.15) that if $s > 0$, then $\mathscr{H}^s(A) = 0$ and, similarly, that $\mathscr{H}^s(B) = 0$.

Let proj denote orthogonal projection from the plane onto L, the line $y = x$. Then $\text{proj}(x, y)$ is the point of L at (signed) distance $2^{-\frac{1}{2}}(x + y)$ from the origin. If $u \in [0, 1]$ we may find $x \in A$ and $y \in B$ such that $u = x + y$ (some of the decimal digits of u are provided by x, the rest by y). Thus $\text{proj}(A \times B)$ is a subinterval of L of length $2^{-\frac{1}{2}}$. Using the fact that orthogonal projection does not increase distances and so, by Lemma 1.8, does not increase Hausdorff measures,

$$2^{-\frac{1}{2}} = \mathscr{H}^1(\text{proj}(A \times B)) \leq \mathscr{H}^1(A \times B).$$

If desired, A and B may be made into compact sets by the addition of countable sets of points. $\quad \square$

All the results of this section may be enunciated in a much more general context; see Larman (1967a) and Wagmann (1969a, 1969b, 1971a) who consider products of sets in very general spaces. In particular, the higher-dimensional analogue of Corollary 5.9, which may be proved in a similar way, is:

Theorem 5.12
Let $A \subset \mathbb{R}^n$ and $B \subset \mathbb{R}^m$. Then

$$\mathcal{H}^{s+t}(A \times B) \geq b \mathcal{H}^s(A) \mathcal{H}^t(B)$$

for some constant b, where $A \times B \subset \mathbb{R}^{n+m}$.

A *cylinder set* is a subset of $\mathbb{R}^3$ of the form $E \times I$, where E is a 1-dimensional subset of $\mathbb{R}^2$ and $I \subset \mathbb{R}$ is the unit interval. A major problem, proposed by Randolph (1936), is to determine when

$$\mathcal{H}^1(E) = \tfrac{1}{4}\pi \mathcal{H}^2(E \times I),$$

that is, when 'area = base × height'. The interested reader is referred to Besicovitch & Moran (1945), Freilich (1965), Ward (1967) and Larman (1967c) for further details.

Exercises on Chapter 5

5.1   Let $E \subset \mathbb{R}$ be an s-set and I the unit interval. Show that $E \times I$ is an $(s+1)$-set in $\mathbb{R}^2$.

5.2   With notation as in Theorem 5.8 show that if E is a Borel set in $\mathbb{R}^2$, then

$$\mathcal{H}^{s+t}(E) \geq b \int \mathcal{H}^t(E_x) \, d\mathcal{H}^s(x).$$

5.3   Let $E \subset \mathbb{R}$ be an s-set formed by a Cantor-like process, with $E = \bigcap_1^\infty E_j$, where each E_j has a δ-cover of equal intervals $\{U_i\}$ such that $\sum |U_i|^s \leq c$, with $\delta \to 0$ as $j \to \infty$. Let $F = \{x - y : x, y \in E\}$ be the difference set for E. Show that $\dim F \leq 2 \dim E$. (Hint: Consider the projection of $E \times E$ onto the line $x + y = 0$.)

5.4   Deduce from Exercise 1.7 that the set

$$G = \{(x, y) : y - x \text{ is irrational}\} \subset \mathbb{R}^2$$

has full plane Lebesgue measure, but contains no subset $A \times B$, where A and B are both Lebesgue measurable sets of positive measure.

6

Projection properties

6.1 Introduction

In this chapter we establish results of the following nature: if E is an s-set in the plane, then the orthogonal projection of E onto lines in almost all directions has dimension t, where $t = s$ if $s \leq 1$, and $t = 1$ if $s \geq 1$. More generally, if Π is a k-dimensional subspace of $\mathbb{R}^n$ and if proj_Π denotes orthogonal projection from $\mathbb{R}^n$ onto Π we investigate the Hausdorff measure of $\text{proj}_\Pi E$ in terms of that of E. The following lemma gives the obvious inequality in one direction.

Lemma 6.1

Let E be any subset of $\mathbb{R}^n$ and let Π be any subspace. Then

$$\mathscr{H}^s(\text{proj}_\Pi E) \leq \mathscr{H}^s(E).$$

Proof. The projection mapping does not increase distances, i.e. $|\text{proj}_\Pi(x) - \text{proj}_\Pi(y)| \leq |x - y|$ $(x, y \in E)$, so the lemma is a direct consequence of Lemma 1.8. $\qquad \square$

The major part of this chapter is concerned with estimates in the opposite sense. The work divides naturally into a general case and a special case. For a general s-set E we depart from the original geometric proofs of Marstrand (1954a) and adopt an approach involving capacities of sets. In the special case where E is an s-set with s integral, the results are delicately balanced; strikingly different phenomena occur when E is regular and when E is irregular, and here we follow the geometrical approach of Besicovitch (1939).

6.2 Hausdorff measure and capacity

This section relates two apparently very different ideas in analysis, as well as paving the way to studying orthogonal projection of s-sets.

The idea of the capacity of a set was originally developed for treating electrostatic problems. The theory was extended to more general laws of attraction in the branch of mathematics known as potential theory, much of the early work being formulated in the famous thesis of Frostman (1935). (For more recent accounts see Taylor (1961), Carleson (1967), Hayman & Kennedy (1976) or Hille (1973).) It turns out that the Hausdorff dimension and the capacity of a set are related, and it is sometimes more convenient to use the latter concept when studying dimensional properties.

The *support* of a Borel measure μ, sometimes written supp μ, is defined as the smallest closed set S such that $\int f \, \mathrm{d}\mu = 0$ for every continuous function f that vanishes on S; intuitively the support may be thought of as the set on which μ is concentrated.

We quote for future reference a suitable version of the well-known Riesz representation theorem which identifies measures with linear functionals. (A complete treatment is given in Kingman & Taylor (1966, Section 9.5), or Rudin (1970, Chapter 2).) Let S be a compact subset of $\mathbb{R}^n$. If $\mathscr{F}$ is the space of continuous functions on S we say that $\psi : \mathscr{F} \to \mathbb{R}$ is a *positive linear functional on $\mathscr{F}$* if

$$\psi(\lambda_1 f_1 + \lambda_2 f_2) = \lambda_1 \psi(f_1) + \lambda_2 \psi(f_2) \quad (\lambda_1, \lambda_2 \in \mathbb{R}, f_1, f_2 \in \mathscr{F})$$

and

$$\psi(f) \geq 0 \quad \text{if } f(x) \geq 0 \text{ for all } x \in S.$$

Clearly, the mapping $\psi(f) = \int_S f \, \mathrm{d}\mu$ is a positive linear functional on $\mathscr{F}$ for any finite Borel measure μ on S. The Riesz representation theorem tells us that the converse is also true.

Theorem 6.2 (Riesz representation theorem)
Let S be a compact subset of $\mathbb{R}^n$ and let ψ be a positive linear functional on the space $\mathscr{F}$ of continuous functions on S. Then there exists a Borel measure μ supported by S with $\mu(S) < \infty$ such that

$$\psi(f) = \int_S f \, \mathrm{d}\mu$$

for all $f \in \mathscr{F}$.

A Borel measure μ on $\mathbb{R}^n$, of compact support and with $0 < \mu(\mathbb{R}^n) < \infty$, is called a *mass distribution*. The *t-potential* at a point x due to the mass distribution μ is defined as

$$\phi_t(x) = \int \frac{\mathrm{d}\mu(y)}{|x - y|^t}.$$

The *t-energy* of μ is given by

$$I_t(\mu) = \int \phi_t(x) \mathrm{d}\mu(x) = \int \int \frac{\mathrm{d}\mu(x)\mathrm{d}\mu(y)}{|x - y|^t}. \tag{6.1}$$

If E is a compact subset of $\mathbb{R}^n$ the *t-capacity* of E, written as $C_t(E)$, is defined by

$$C_t(E) = \sup_{\mu} \left\{ \frac{1}{I_t(\mu)} : E \text{ supports } \mu \text{ and } \mu(E) = 1 \right\}. \qquad (6.2)$$

(Note that potentials and energies may be infinite; we adopt the convention that $1/\infty = 0$.) For an arbitrary subset E of $\mathbb{R}^n$, define

$$C_t(E) = \sup \{ C_t(F) : F \text{ is compact}, F \subset E \}. \qquad (6.3)$$

Some authors give slightly different definitions of capacity, though such definitions are equivalent in the important sense that the same sets have capacity zero. A basic result of potential theory is that (6.2) is equivalent to

$$C_t(E) = \sup_{\mu : \mu(E) = 1} \left(\sup_{x \in \text{supp} \mu} \{ t\text{-potential of } \mu \text{ at } x \} \right)^{-1}.$$

Moreover, these various bounds are attained. We do not require such results here; the details are contained in, for example, Hayman & Kennedy (1976).

It is apparent that if $E \subset E'$, then $C_t(E) \le C_t(E')$ for any t. Also, if μ is any mass distribution with $I_t(\mu) = \infty$, then $I_s(\mu) = \infty$ for any $s > t$. Thus if $C_t(E) = 0$, then $C_s(E) = 0$ if $s > t$.

We note from the definitions the fundamental fact that $C_t(E) > 0$ if and only if there exists a mass distribution μ with support contained in E such that $I_t(\mu) < \infty$.

Lemma 6.3

Let E be a compact subset of $\mathbb{R}^n$ with $\mathcal{H}^s(E) < \infty$. Let μ be a mass distribution supported by E and let

$$F_0 = \{ x \in E : \overline{\lim_{r \to 0}} \, \mu(B_r(x)) / r^s = 0 \}.$$

Then $\mu(F_0) = 0$.

Proof. Fix $\alpha, \rho > 0$ and let

$$F = \{ x \in E : \mu(B_r(x)) / r^s < \alpha \text{ for all } r \le \rho \}.$$

If $\{U_i\}$ is any δ-cover of F with $\delta \le \rho$, then, assuming that each U_i contains some point of F, there exist balls $\{B_i\}$ centred in F with $|B_i| \le 2|U_i| \le 2\rho$ and with $U_i \subset B_i$ for each i. Thus

$$\mu(F) \le \sum_i \mu(B_i) < 2^{-s}\alpha \sum_i |B_i|^s \le \alpha \sum_i |U_i|^s.$$

This holds for any δ-cover $\{U_i\}$, so $\mu(F) \le \alpha \mathcal{H}^s_\delta(F)$, giving $\mu(F) \le \alpha \mathcal{H}^s(F) \le \alpha \mathcal{H}^s(E)$. Since ρ and α may be chosen arbitrarily small, it follows that $\mu(F_0) = 0$. $\square$

We now obtain the basic relationship between Hausdorff measures and capacities. Part (a) of Theorem 6.4 was proved by Frostman (1935) using an early version of a comparable net measure, see also Wallin (1969). Part (b), due to Erdös & Gillis (1937), has been further generalized by Ugaheri (1942) and Kametani (1945).

Theorem 6.4
Let E be a subset of $\mathbb{R}^n$.
(a) *If E is a Souslin set with $C_t(E) = 0$, then $\mathscr{H}^s(E) = 0$ for all $s > t$.*
(b) *If $\mathscr{H}^s(E) < \infty$, then $C_s(E) = 0$.*

Proof. (a) Suppose that E is a Souslin set with $\mathscr{H}^s(E) > 0$. We show that $C_t(E) > 0$ if $t < s$ by producing a mass distribution μ supported by a compact subset of E such that $I_t(\mu) < \infty$.

By Theorem 5.6(b) there exists a compact set $F \subset E$ with $0 < \mathscr{H}^s(F) < \infty$ such that

$$\mathscr{H}^s(B_r(x) \cap F) \leq br^s \qquad (x \in \mathbb{R}^n, r < 1)$$

for some constant b. Let μ be the restriction of $\mathscr{H}^s$ to F, so that μ is a mass distribution supported by F. Take $x \in \mathbb{R}^n$ and let

$$m(r) = \mu(B_r(x)) = \mathscr{H}^s(B_r(x) \cap F) \leq br^s \quad (r \leq 1). \tag{6.4}$$

Then

$$\phi_t(x) = \int_{|x-y| \leq 1} \frac{\mathrm{d}\mu(y)}{|x - y|^t} + \int_{|x-y| > 1} \frac{\mathrm{d}\mu(y)}{|x - y|^t}$$

$$\leq \int_0^1 r^{-t} \mathrm{d}m(r) + \mu(\mathbb{R}^n)$$

$$= [r^{-t} m(r)]_{0+}^1 + t \int_0^1 r^{-(t+1)} m(r) \mathrm{d}r + \mu(\mathbb{R}^n)$$

$$\leq b + bt \int_0^1 r^{s-t-1} \mathrm{d}r + \mu(\mathbb{R}^n)$$

$$= b(1 + t/(s - t)) + \mathscr{H}^s(F),$$

after integrating by parts and using (6.4). Thus $\phi_t(x)$ is uniformly bounded on $\mathbb{R}^n$, so

$$I_t(\mu) = \int \phi_t(x) \mathrm{d}\mu(x) < \infty,$$

as required.
(b) By (6.3) it is enough to prove the result when E is a compact set. Let μ be any mass distribution supported by E; we show that $I_s(\mu) = \infty$. Let E_0 be the set

$$E_0 = \{x \in E : \overline{\lim_{r \to 0}} \, \mu(B_r(x))/r^s > 0\}.$$

If $x \in E_0$ we may find a sequence $\{r_i\}$ decreasing to 0 such that

$$\mu(B_{r_i}(x)) \geq \varepsilon r_i^s$$

for some $\varepsilon > 0$. Unless $\mu(\{x\}) > 0$ (in which case it is evident that $I_s(\mu) = \infty$) it follows from the continuity of μ that there exists q_i with $0 < q_i < r_i$ and

$$\mu(A_i) \geq \tfrac{1}{2}\varepsilon r_i^s \qquad (i = 1, 2, \ldots),$$

where A_i is the annular region $B_{r_i}(x) \backslash B_{q_i}(x)$. Taking subsequences if necessary, we may assume that $r_{i+1} < q_i$ for all i, making the $\{A_i\}$ disjoint. Hence if $x \in E_0$,

$$\phi_s(x) = \int \frac{\mathrm{d}\mu(y)}{|x - y|^s} \geq \sum_i \int_{A_i} \frac{\mathrm{d}\mu(y)}{|x - y|^s}$$

$$\geq \sum_{i=1}^{\infty} \tfrac{1}{2}\varepsilon r_i^s r_i^{-s} = \infty.$$

By Lemma 6.3 the Borel set E_0 contains μ-almost all of the points of E, so $\mu(E_0) > 0$ and

$$I_s(\mu) = \int \phi_s(x) \mathrm{d}\mu(x) = \infty.$$

As this is true for any mass distribution μ supported by E we conclude that $C_s(E) = 0$. $\qquad \square$

Corollary 6.5
If E is a Souslin subset of $\mathbb{R}^n$, then

$$\dim E = \inf\{t : C_t(E) = 0\} = \sup\{t : C_t(E) > 0\}.$$

Proof. As $\mathcal{H}^s(E) = \infty$ for $s < \dim E$ and $\mathcal{H}^s(E) = 0$ for $s > \dim E$ (see (1.14)), the corollary follows immediately. $\qquad \square$

Corollary 6.5 is often used in the following form:

Corollary 6.6
Let E be a Souslin subset of $\mathbb{R}^n$.
(a) If $I_t(\mu) < \infty$ for some mass distribution μ supported by E, then $t \leq \dim E$.
(b) If $t < \dim E$, then there exists a mass distribution μ with support in E such that $I_t(\mu) < \infty$.

Note that Hausdorff measure and capacity are not completely equivalent concepts. If $\dim E = t$ there are various possibilities for the relative values of $C_t(E)$ and $\mathcal{H}^t(E)$; see Garnett (1970) or Mattila (1984c) for some examples.

Nowadays, Fourier transforms play a vital part in potential theory (see Rudin (1973) for a general treatment of transform theory). Here we merely require the definition, and a consequence of Plancherel's result on transforms of square-integrable functions. If f and μ are, respectively, an integrable function and a mass distribution on $\mathbb{R}$, their (1-dimensional) Fourier transforms are given by

$$\left.\begin{aligned}
\hat{f}(p) &= (2\pi)^{-\frac{1}{2}} \int f(u)e^{iup}\mathrm{d}u, \\[2ex]
\hat{\mu}(p) &= (2\pi)^{-\frac{1}{2}} \int e^{iup}\mathrm{d}\mu(u).
\end{aligned}\right\} \qquad (p\in\mathbb{R})$$

An elementary result (the Riemann–Lebesgue lemma) asserts that $\hat{f}$ and $\hat{\mu}$ are bounded continuous functions

Lemma 6.7
Let μ be a mass distribution on $\mathbb{R}$ such that

$$0 < \int |\hat{\mu}(p)|^2 \mathrm{d}p < \infty.$$

Then the support of μ has positive Lebesgue measure.

Proof. Plancherel's theorem states that square-integrable Fourier transforms correspond to square-integrable *functions*. Hence μ is a Borel measure such that, for any continuous function f that vanishes outside some bounded interval,

$$\int f(u)\mathrm{d}\mu(u) = \int f(u)g(u)\mathrm{d}u, \qquad (6.5)$$

where g is a Borel-measurable function with

$$0 < \int |\hat{\mu}(p)|^2 \mathrm{d}p = \int |g(u)|^2\mathrm{d}u. \qquad (6.6)$$

If μ were supported by a compact set of zero Lebesgue measure, (6.5) would imply that $g(u) = 0$ for almost all u, which is impossible by (6.6). $\qquad\square$

6.3 Projection properties of sets of arbitrary dimension

We now prove the most important result on the projection of sets of general dimension. Results of this kind were first proved directly (without the use of capacities) by Marstrand (1954a). The potential-theoretic proof given here is essentially that of Kaufman (1968). So the basic ideas do not become too clouded by notation, we give the proof in two dimensions and mention generalizations later. In this case proj_θ denotes orthogonal projection from $\mathbb{R}^2$ onto L_θ, the line through the origin making angle θ with some fixed axis. 1-dimensional Lebesgue measure on subsets of L_θ will be denoted by $\mathscr{L}^1$ in the obvious way. Here, 'almost all θ' refers to Lebesgue measure.

Theorem 6.8

Let E be a Souslin subset of $\mathbb{R}^2$ with $\dim E = s$.
(a) *If $s \leq 1$, then $\dim(\text{proj}_\theta E) = s$ for almost all $\theta \in [0, \pi)$.*
(b) *If $s > 1$, then $\mathcal{L}^1(\text{proj}_\theta E) > 0$ for almost all $\theta \in [0, \pi)$.*

Proof. (a) By Lemma 6.1, $\dim(\text{proj}_\theta E) \leq s$ for all θ. If $t < s < 1$ we may, using Corollary 6.6(b), choose a mass distribution μ, with support contained in a compact subset of E, such that $I_t(\mu) < \infty$. The mapping $f \mapsto \int f(x \cdot \theta) d\mu(x)$ is a positive linear functional, so, by the Riesz representation theorem, Theorem 6.2, we may find a mass distribution μ_θ on $\mathbb{R}$ for each θ, such that

$$\int f(u) d\mu_\theta(u) = \int f(x \cdot \theta) d\mu(x) \tag{6.7}$$

for continuous functions f. (Here θ denotes the unit vector in direction θ and '$\cdot$' is the usual scalar product, with x identified with the vector from the origin to x.) By the usual approximation process using the monotone convergence theorem for integrals, (6.7) holds for any non-negative measurable function. Identifying $u \in \mathbb{R}$ with the point $u\theta$ on L_θ we may regard μ_θ as a mass distribution on L_θ, with support contained in $\text{proj}_\theta E$. Now,

$$I_t(\mu_\theta) = \int \int \frac{d\mu_\theta(u) d\mu_\theta(v)}{|u - v|^t}$$

$$= \int \int \frac{d\mu(x) d\mu(y)}{|x \cdot \theta - y \cdot \theta|^t}$$

$$= \int \int \frac{d\mu(x) d\mu(y)}{|(x - y) \cdot \theta|^t}.$$

Applying Fubini's theorem to $[0, \pi) \times \mathbb{R}^2 \times \mathbb{R}^2$, we see that $I_t(\mu_\theta)$ is a measurable function of θ and that

$$\int_0^\pi I_t(\mu_\theta) d\theta = c \int \int \frac{d\mu(x) d\mu(y)}{|x - y|^t} = c I_t(\mu) < \infty,$$

where $c = \int_0^\pi |\theta \cdot \tau|^{-t} d\theta < \infty$ (since $t < 1$ and this integral is independent of the unit vector τ). Thus $I_t(\mu_\theta) < \infty$ for almost all θ, so we conclude from Corollary 6.6(a) that $\dim(\text{proj}_\theta E) \geq t$. This is true for all $t < s$, so $\dim(\text{proj}_\theta E) = s$ for almost all θ.
(b) If $s > 1$, then there exists a mass distribution μ supported by E with $I_1(\mu) < \infty$, by Corollary 6.6(b). As in the proof of (a) we define mass distributions μ_θ on L_θ by (6.7), but on this occasion we examine the Fourier transforms $\hat{\mu}_\theta$

of the measures μ_θ.

$$|\hat\mu_\theta(p)|^2 = \frac{1}{2\pi}\int e^{iup}d\mu_\theta(u)\int e^{-ivp}d\mu_\theta(v)$$

$$= \frac{1}{2\pi}\int\int e^{i(u-v)p}d\mu_\theta(u)d\mu_\theta(v) = \frac{1}{2\pi}\int\int e^{ip(x-y)\cdot\theta}d\mu(x)d\mu(y),$$

using (6.7). Then

$$|\hat\mu_\theta(p)|^2 + |\hat\mu_{\theta+\pi}(p)|^2 = \frac{1}{\pi}\int\int\cos(p(x-y)\cdot\theta)d\mu(x)d\mu(y),$$

so that

$$\int_0^{2\pi}|\hat\mu_\theta(p)|^2d\theta = \frac{1}{2\pi}\int_0^{2\pi}\int\int\cos(p|x-y|\cos\theta)d\mu(x)d\mu(y)d\theta,$$

since the definite integral is again independent of the argument of $x-y$.
Thus, using Fubini's theorem and the definition of the Bessel function,
$J_0(u) = (2\pi)^{-1}\int_0^{2\pi}\cos(u\cos\theta)d\theta$, we obtain

$$\int_0^{2\pi}|\hat\mu_\theta(p)|^2d\theta = \int\int J_0(p|x-y|)d\mu(x)d\mu(y).$$

(The detailed theory of Bessel functions is described in Watson (1966). If
$0 < m < \infty$ Fubini's theorem gives

$$\int_{-m}^m\int_0^{2\pi}|\hat\mu_\theta(p)|^2d\theta dp = \int\int\int_{-m}^m J_0(p|x-y|)dp\,d\mu(x)d\mu(y)$$

$$= \int\int\int_{u=-m|x-y|}^{m|x-y|} J_0(u)\frac{1}{|x-y|}du\,d\mu(x)d\mu(y)$$

$$\le b\int\int\frac{d\mu(x)d\mu(y)}{|x-y|} = bI_1(\mu),$$

where b is independent of m. (The improper Riemann integral $\int_{-\infty}^\infty J_0(u)du$
is convergent so that the definite integrals $\int_{-m}^m J_0(u)du$ are uniformly
bounded as m varies.)

Letting $m\to\infty$ and again using Fubini's theorem (it is easy to see that
$\hat\mu_\theta(p)$ is continuous in both θ and p),

$$\int_0^{2\pi}\int_{-\infty}^\infty|\hat\mu_\theta(p)|^2dp\,d\theta \le bI_1(\mu) < \infty,$$

so that $\int_{-\infty}^\infty|\hat\mu_\theta(p)|^2dp < \infty$ for almost all θ in $[0,\pi)$. By Lemma 6.7 the
support of μ_θ has positive Lebesgue measure for any such θ. But the
support of μ_θ is just the projection of the support of μ (the projection of a
compact set being compact) and this is contained in the Souslin set $\text{proj}_\theta E$.

Hence $\text{proj}_\theta E$ is measurable and has positive Lebesgue measure for almost all θ.　□

Theorem 6.8 has various higher-dimensional analogues depending on the dimension of the subspaces of projection. Denote by $G_{n,k}$ the Grassmann manifold consisting of all k-dimensional subspaces of $\mathbb{R}^n$. $G_{n,k}$ may be endowed with a normalized rotation invariant measure in a natural way, and when we speak of 'almost all $\Pi \in G_{n,k}$' we refer to this measure.

Kaufman's method, using higher-dimensional Fourier transforms, and the method of Marstrand (1954*a*) generalize without difficulty, see Mattila (1975*b*).

Theorem 6.9
Let E be a Souslin subset of $\mathbb{R}^n$ with $\dim E = s$.
(a) If $s \leq k$, then $\dim(\text{proj}_\Pi E) = s$ for almost all $\Pi \in G_{n,k}$.
(b) If $s > k$, then $\text{proj}_\Pi E$ has positive k-dimensional Lebesgue measure for almost all $\Pi \in G_{n,k}$.

An alternative method of obtaining such theorems is to integrate the Fourier transforms of measures projected onto subspaces, using the important expression for the energy of a measure μ in terms of its n-dimensional Fourier transform,

$$I_t(\mu) = c \int_{\mathbb{R}^n} \frac{|\hat{\mu}(p)|^2}{|p|^{n-t}} \, dp, \tag{6.8}$$

where c depends only on n and t. Formally this equality is a consequence of the convolution theorem in transform theory, but a rigorous derivation requires some care (see Carleson (1967, p. 23)). Falconer (1982) uses such a method to obtain non-trivial upper bounds on the Hausdorff dimension of the exceptional sets in Theorems 6.8 and 6.9. (The exceptional sets are the sets of θ, respectively Π, for which the conclusion of the theorems fail.) Kaufman's original proof and Mattila's (1975*b*) generalization provide other information on the size of the exceptional sets. The paper by Kaufman & Mattila (1975) gives examples of sets in $\mathbb{R}^n$ for which the exceptional set is as large as possible, see Theorem 8.17.

Recent work of Mattila (1975*b*) and Falconer (1982) is couched in terms of capacities rather than Hausdorff measures and is consequently slightly more general than that above.

Marstrand's (1954*a*) original enunciation of the projection theorem includes a variation which is sometimes useful, and which may be adapted to any of the results discussed. If E is an s-set in $\mathbb{R}^2$ with $s > 1$, then, for almost all θ, not only does $\text{proj}_\theta E$ have positive Lebesgue measure but also, if F is any Souslin subset of E with $\mathscr{H}^s(F) > 0$, then $\mathrm{proj}_\theta F$ has positive Lebesgue measure. In other words the exceptional set of θ does not depend on the subset F chosen.

Mattila's (1975b) version of Theorem 6.9(b) has the stronger conclusion that if $\dim E > k$, then $\int_{u \in \Pi} \mathscr{H}^0(E \cap \mathrm{proj}_\Pi^{-j} u) \mathrm{d}u = \infty$ for almost all $\Pi \in G_{n,k}$ ($\mathscr{H}^0$ simply counts the number of points in the set). A further consequence of Mattila's paper is that if $0 < t < \dim E - k$, the set of $u \in \Pi$ for which $\mathscr{H}^t(E \cap \mathrm{proj}_\Pi^{-j} u) = \infty$ has positive $\mathscr{H}^k$-measure for almost all $\Pi \in G_{n,k}$.

Davies (1979, 1982) has recently shown, using the continuum hypothesis, that these results fail for sets that are merely 'essentially s-dimensional'. He constructs a set that cannot be expressed as a countable union of sets of dimension less than 2, but with projection onto almost every line of Lebesgue measure zero.

6.4 Projection properties of sets of integral dimension

This section, which in a sense follows on from Chapter 3, concerns the projections of an s-set E when s is an integer. We restrict our proofs to the case where E is a linearly measurable subset of the plane; analysis of the higher-dimensional analogues is much more complicated.

There is a stark contrast between the projection behaviour of regular and irregular sets of integral dimension. A regular 1-set in $\mathbb{R}^2$ projects to a set of positive 1-dimensional measure in almost all directions, whereas an irregular 1-set projects to a set of zero 1-dimensional measure in almost all directions.

Regular sets are easy to deal with. As before, proj_θ denotes orthogonal projection onto the line L_θ that makes angle θ with some axis.

Theorem 6.10
Let E be a regular 1-set in $\mathbb{R}^2$. Then $\mathscr{L}^1(\mathrm{proj}_\theta E) > 0$ for all but at most one value of θ.

Proof. By Theorem 3.25 a regular 1-set comprises a measurable subset of a countable union of rectifiable curves and a set of measure zero, so it is enough to prove the result if E is a 1-set contained in a rectifiable curve Γ.

Let x be a point of E that is a regular point of both E and Γ. Then, given $\varepsilon > 0$, we may find r such that both

$$\mathscr{H}^1(E \cap B_r(x)) > (1 - \varepsilon^2)2r \quad \text{and} \quad \mathscr{H}^1(\Gamma \cap B_r(x)) < (1 + \varepsilon)2r,$$

implying

$$\mathscr{H}^1(E \cap B_r(x)) > (1 - \varepsilon)\mathscr{H}^1(\Gamma \cap B_r(x)),$$

and hence

$$\mathscr{H}^1((\Gamma \backslash E) \cap B_r(x)) < \varepsilon \mathscr{H}^1(\Gamma \cap B_r(x)).$$

Since $\Gamma \cap B_r(x)$ consists of, at most, a countable number of disjoint arcs, and since Γ is rectifiable, we may choose an arc $\Gamma_0 \subset \Gamma \cap B_r(x)$ such that

$$\mathcal{H}^1(\Gamma_0 \backslash E) < \varepsilon \mathcal{H}^1(\Gamma_0) < 2\varepsilon |y - z|,$$

where y and z are the endpoints of Γ_0. Let L_θ be a line making an angle of ϕ with $[y, z]$, where $\cos \phi > 2\varepsilon$. Then, projecting onto L_θ and using Lemma 6.1,

$$\mathcal{L}^1(\text{proj}_\theta E) > |y - z| \cos \phi - \mathcal{L}^1(\text{proj}_\theta(\Gamma_0 \backslash E))$$
$$\geq |y - z| \cos \phi - \mathcal{H}^1(\Gamma_0 \backslash E) > |y - z|(\cos \phi - 2\varepsilon) > 0.$$

Thus $\mathcal{L}^1(\text{proj}_\theta E) > 0$, except for a set of directions θ contained in an interval of length $2 \sin^{-1} 2\varepsilon$ for all $\varepsilon > 0$, that is, except for, at most, one direction. $\quad\square$

It is possible to prove results of a more quantitative nature than the above. For example, the $\mathcal{H}^1$-measure of a regular 1-set E in $\mathbb{R}^2$ may be found by integrating the number of points in $E \cap L$ over all lines L in the plane, where the set of lines is endowed with a suitable invariant measure. Ideas such as this form the basis of integral geometry; the interested reader should consult Santaló (1976) or Federer (1969) for further details.

We turn to the projection of irregular 1-sets in the plane. Gillis (1934a, 1936a) showed that at almost all x in an irregular 1-set E, $\lim_{r \to 0} \mathcal{L}^1(\text{proj}_\theta(E \cap B_r(x))/\mathcal{H}^1(E \cap B_r(x)) = 0$ for almost all θ, but Besicovitch (1939) proved the natural result that $\mathcal{L}^1(\text{proj}_\theta E) = 0$ for almost all θ. The starting point of the proof is the fact that an irregular 1-set has tangents almost nowhere (Corollary 3.30). Throughout the proof I's and J's will denote intervals of directions in $[0, \pi)$. Following Besicovitch we start with three definitions.

A direction θ is a *condensation direction of the first kind* at the point $x \in E$ if $L_\theta(x)$, the line through x in direction θ, intersects E infinitely often in every neighbourhood of x.

Given $x \in E$ and positive numbers ρ, ε and m we define a subset $T(x, \rho, \varepsilon, m)$ of $[0, \pi)$ by taking $\theta \in T(x, \rho, \varepsilon, m)$ if and only if there exists r with $0 < r < \rho$ and some open interval I with $\theta \in I \subset [0, \pi)$ and $|I| < \varepsilon$ for which

$$\frac{\mathcal{H}^1(E \cap C_r(x, I))}{r} \geq m|I|. \tag{6.9}$$

($C_r(x, I)$ denotes the double sector consisting of the points of $B_r(x)$ that lie on $L_\theta(x)$ for some $\theta \in I$.) Observe that if E is irregular, $\mathcal{H}^1(E \cap C_r(x, I))$ is continuous in r, from which it is easy to see that $T(x, \rho, \varepsilon, m)$ is a Lebesgue measurable subset of $[0, \pi)$.

A direction θ is a *condensation direction of the second kind* at the point

$x \in E$ if $\theta \in T$, where

$$T = \bigcap_{\rho > 0} \bigcap_{\varepsilon > 0} \bigcap_{0 < m < \infty} T(x, \rho, \varepsilon, m). \tag{6.10}$$

A point x is called a *point of radiation of E* if almost all θ in $[0, \pi)$ (in the sense of Lebesgue measure) are condensation directions of E at x.

The crux of the proof of Besicovitch's theorem on the projection of irregular sets is contained in the next lemma, which depends on an ingenious application of Lebesgue's density theorem.

Lemma 6.11

Let E be a closed irregular 1-set in the plane. Then almost all points of E are points of radiation.

Proof. We know from Corollary 3.30 that for $\mathcal{H}^1$-almost all $x \in E$ and for any interval $I \subset [0, \pi)$,

$$\overline{\lim_{r \to 0}} \frac{\mathcal{H}^1(E \cap C_r(x, I))}{2r} > \tfrac{1}{20}|I|. \tag{6.11}$$

Fix x as such a point; we show that x is a point of radiation of E.

First note that the set S of condensation directions of the first kind at x is a Lebesgue-measurable subset of $[0, \pi)$. For if $S(r)$ is the set of θ for which $L_\theta(x) \cap B_r(x)$ contains points of E other than x, then $S(r)$ is the union of the closed sets $\{\theta : \text{there exists } y \in L_\theta \text{ with } \delta \le |x - y| \le r\}$ over positive rational values of δ, making $S(r)$ a Borel set. But $S = \bigcap_{j=1}^{\infty} S(1/j)$, so S is also a Borel set and is thus Lebesgue measurable.

We aim to show that if ρ, ε and m are positive, then $\theta \in T(x, \rho, \varepsilon, m)$ for almost all $\theta \notin S$. Let $\theta \in [0, \pi)$ be a direction at which the Lebesgue density of S, $\lim_{r \to 0} \mathcal{L}^1(S \cap [\theta - r, \theta + r])/2r$, is zero. (In this context $\mathcal{L}^1$ is Lebesgue measure on $[0, \pi)$.) By Lebesgue's theorem, Theorem 1.13, this is true for almost all $\theta \notin S$. Then, for all sufficiently small intervals I with $\theta \in I \subset [0, \pi)$,

$$\mathcal{L}^1(S \cap I) < \frac{|I|}{20m}. \tag{6.12}$$

Let I be any such interval with $0 < |I| < \varepsilon$. As $S(r) \searrow S$ as $r \to 0$, there exists $\rho_1 \le \rho$ such that if $r < \rho_1$, then

$$\mathcal{L}^1(S(r) \cap I) < \frac{|I|}{20m}. \tag{6.13}$$

Using (6.11) choose $r < \rho_1$ such that

$$\mathcal{H}^1(E \cap C_r(x, I)) > \frac{2r|I|}{20}. \tag{6.14}$$

If it should happen that

$$\mathscr{H}^1(E \cap C_r(x, I)) > mr|I|, \tag{6.15}$$

then, by definition, $\theta \in T(x, \rho, \varepsilon, m)$. Otherwise, by (6.13) and the definition of Lebesgue measure, there exists a countable collection of disjoint intervals $\{I_i\}$ such that

$$S(r) \cap I \subset \bigcup_1^\infty I_i \tag{6.16}$$

and

$$\mathscr{L}^1(\bigcup_1^\infty I_i) < \frac{|I|}{20\,m}. \tag{6.17}$$

Let Q be the set of indices i for which

$$\mathscr{H}^1(E \cap C_r(x, I_i)) > mr|I_i|. \tag{6.18}$$

By (6.17),

$$\sum_{i \notin Q} \mathscr{H}^1(E \cap C_r(x, I_i)) \le mr\mathscr{L}^1(\bigcup_{i \notin Q} I_i) < \frac{r|I|}{20}. \tag{6.19}$$

But the points of $E \cap C_r(x, I)$ all lie on lines $L_\theta(x)$ with $\theta \in S(r) \cap I$, so by (6.16),

$$\bigcup_{i=1}^\infty (E \cap C_r(x, I_i)) \supset E \cap C_r(x, S(r) \cap I) \supset E \cap C_r(x, I).$$

Hence

$$\sum_{i \in Q} \mathscr{H}^1(E \cap C_r(x, I_i)) = \sum_{i=1}^\infty \mathscr{H}^1(E \cap C_r(x, I_i))$$

$$- \sum_{i \notin Q} \mathscr{H}^1(E \cap C_r(x, I_i))$$

$$> \mathscr{H}^1(E \cap C_r(x, I)) - \frac{r|I|}{20} > \frac{r|I|}{20}, \tag{6.20}$$

by (6.19) and (6.14). It follows using (6.18) that by a process of expanding and combining the intervals $\{I_i\}_{i \in Q}$ we may obtain a disjoint collection of open intervals $\{J_j\}_1^\infty$ with

$$J = \bigcup_{j=1}^\infty J_j \supset \bigcup_{i \in Q} I_i$$

and

$$\mathscr{H}^1(E \cap C_r(x, J_j)) = mr|J_j| \quad (j = 1, 2, \ldots); \tag{6.21}$$

since (6.15) is assumed not to hold, this may be done in such a way that

$J \subset I$. From (6.20) and (6.21),

$$\mathscr{L}^1(J) = \sum_j |J_j| > \frac{|I|}{20m}.$$

If $\theta' \in J$, then $\theta' \in J_j$ for some j, so, by (6.21), $\theta' \in T(x, \rho, \varepsilon, m)$. Thus $J \subset T(x, \rho, \varepsilon, m) \cap I$, so

$$\mathscr{L}^1(T(x, \rho, \varepsilon, m) \cap I) > \frac{|I|}{20m}.$$

This holds for all sufficiently small intervals I containing θ, so we have shown that for almost all $\theta \notin S$, either $\theta \in T(x, \rho, \varepsilon, m)$, or the lower Lebesgue density of $T(x, \rho, \varepsilon, m)$ at θ is at least $1/(20m)$. It follows from the Lebesgue density theorem, Theorem 1.13, that the Lebesgue density of $T(x, \rho, \varepsilon, m)$ is zero at almost all points of its complement, so almost all θ not in S belong to $T(x, \rho, \varepsilon, m)$.

This is true for all values of ρ, ε and m so, as the intersections in (6.10) may be taken to be countable (since $T(x, \rho, \varepsilon, m)$ decreases as ρ and ε decrease and as m increases), we conclude that almost all θ not in S are in T. This completes the proof of the lemma. $\square$

By a simple modification to the second paragraph of the proof, Lemma 6.11 holds for any irregular 1-set E, though we do not need to use this.

It is not known whether Lemma 6.11 can be strengthened to show that almost all points of E are points of radiation of the second kind (that is, have almost all directions as condensation directions of the second kind).

Now consider the product measure $\mathscr{H}^1 \times \mathscr{L}^1$ on $E \times [0, \pi)$ defined by the requirement that $(\mathscr{H}^1 \times \mathscr{L}^1)(A \times B) = \mathscr{H}^1(A)\mathscr{L}^1(B)$ for Borel sets A and B. Let G_1 and G_2 be the subsets given by

$$G_i = \{(x, \theta) : \theta \text{ is a condensation direction of the } i\text{th kind of } E \\ \text{at } x\} \quad (i = 1, 2).$$

Lemma 6.12

If E is a closed irregular 1-set, then G_1 and G_2 are Borel subsets of $E \times [0, \pi)$ and are thus $(\mathscr{H}^1 \times \mathscr{L}^1)$-measurable.

Proof. (a) For $0 < r < \rho$ write

$$G_{r,\rho} = \{(x, \theta) : x \in E \text{ and } r \le |x - y| \le \rho \\ \text{for some } y \in E \cap L_\theta(x)\}.$$

Then $G_{r,\rho}$ is closed since E is closed. Since $G_{r,\rho}$ increases as r decreases, and decreases as ρ decreases,

$$G_1 = \bigcap_{\rho > 0} [\bigcup_{r \le \rho} G_{r,\rho}]$$

is a Borel set.

(b) As E is irregular, $\mathcal{H}(E \cap C_r(x, I))$ is continuous in x, r and I. Thus if $T'(x, \rho, \varepsilon, m)$ is defined in the same way as $T(x, \rho, \varepsilon, m)$ but with *strict* inequality in (6.9), then $\{(x, \theta) : \theta \in T'(x, \rho, \varepsilon, m)\}$ is open in $E \times [0, \pi)$. But

$$G_2 = \bigcap_{\rho > 0} \bigcap_{\varepsilon > 0} \bigcap_{0 < m < \infty} \{(x, \theta) : \theta \in T'(x, \rho, \varepsilon, m)\}.$$

(We may write T' instead of T here since $T(x, \rho, \varepsilon, m)$ is monotonic in m.) These intersections may be taken over rational values of ρ, ε and m, so G_2 is a Borel set. $\quad\square$

We can now deduce the projection theorem.

Theorem 6.13
Let E be an irregular 1-set in $\mathbb{R}^2$. Then

$$\mathcal{L}^1(\mathrm{proj}_\theta E) = 0 \ \text{for almost all } \theta \in [0, \pi).$$

Proof. First suppose that E is compact. By Lemma 6.12 the set $G = G_1 \cup G_2$ is a $(\mathcal{H}^1 \times \mathcal{L}^1)$-measurable subset of $E \times [0, \pi)$ and by Lemma 6.11

$$\mathcal{L}^1\{\theta : (x, \theta) \in G\} = \pi$$

for almost all $x \in E$. The conditions for Fubini's theorem hold and we conclude that for $\mathcal{L}^1$-almost all $\theta \in [0, \pi)$ we have $(x, \theta) \in G$ for $\mathcal{H}^1$-almost all x in E. Let θ be such a direction so that θ is a condensation direction for almost all x, and let L_ϕ be the line through the origin perpendicular to θ (so that $\phi = \theta + \frac{1}{2}\pi \pmod{\pi}$). We claim that the projection of E onto this line has zero Lebesgue measure.

Let $E = E_0 \cup E_1 \cup E_2$, where $\mathcal{H}^1(E_0) = 0$ and where E_i are the points of E for which θ is a condensation direction of the ith kind for $i = 1, 2$. Then
(a) By Lemma 6.1 $\mathcal{L}^1(\mathrm{proj}_\phi E_0) = 0$.
(b) To deal with E_1 we use Theorem 5.8 on the Cartesian product of sets. If $x \in E_1$, then $L_\theta(x)$ intersects E in infinitely many points, i.e. $\mathcal{H}^0(L_\theta(x) \cap E) = \infty$. Regarding $\mathbb{R}^2$ as the product $L_\phi \times L_\theta$ and applying Theorem 5.8 with $s = 1$ and $t = 0$,

$$\infty > \mathcal{H}^1(E_1) \geq bc\mathcal{H}^1(\mathrm{proj}_\phi E_1)$$

for arbitrarily large c, giving $\mathcal{H}^1(\mathrm{proj}_\phi E_1) = \mathcal{L}^1(\mathrm{proj}_\phi E_1) = 0$.
(c) Let m be any positive number and let $\mathcal{V}$ be the class of subintervals of L_ϕ given by

$$\mathcal{V} = \{J : \mathcal{H}^1(x \in E : \mathrm{proj}_\phi x \in J) > m|J|\}.$$

Then using (6.9) and the definition of E_2 and noting that if $\theta \in I$, then $C_r(x, I)$ may be enclosed in a rectangle of width less than $|I|/r$ with one pair of sides in direction θ, it follows that $\mathcal{V}$ is a Vitali cover for $\mathrm{proj}_\phi E_2$. By Vitali's theorem, Theorem 1.10(b), given $\varepsilon > 0$, there exist disjoint intervals $\{J_i\}$ of $\mathscr{V}$ such that

$$\mathscr{L}^1(\text{proj}_\phi E_2) \le \sum |J_i| + \varepsilon \le \frac{1}{m} \mathscr{H}^1(E) + \varepsilon.$$

Hence, as $1/m$ and ε may be chosen arbitrarily small,

$$\mathscr{L}^1(\text{proj}_\phi E_2) = 0.$$

Taking (a)–(c) together, $\mathscr{L}^1(\text{proj}_\phi E) = 0$, and this is true for almost all ϕ.

Finally, if E is any measurable 1-set, we may, using the regularity of $\mathscr{H}^1$, write $E = F \cup G$, where F is compact and $\mathscr{H}^1(G) < \delta$. We have shown $\mathscr{L}^1(\text{proj}_\theta F) = 0$ for almost all θ, so $\mathscr{L}^1(\text{proj}_\theta E) = \mathscr{L}^1(\text{proj}_\theta G) \le \mathscr{H}^1(G) < \delta$ by Lemma 6.1. But δ may be chosen arbitrarily small, so $\mathscr{H}^1(\text{proj}_\theta E) = 0$ for almost all $\theta \in [0, \pi)$. $\square$

The next corollary provides a very useful test for an irregular 1-set.

Corollary 6.14
A 1-set E in $\mathbb{R}^2$ is irregular if and only if it has projections of Lebesgue measure zero in two distinct directions.

Proof. If E is irregular, then we may certainly find two directions in which the projections have $\mathscr{L}^1$-measure zero, by Theorem 6.13. On the other hand, if E has a regular part of positive linear measure, then it projects to a set of positive $\mathscr{L}^1$-measure by Theorem 6.10. $\square$

In view of Theorem 6.13, one is prompted to ask whether a 1-set can project to a set of measure zero in *all* directions. The following construction of Besicovitch (1928a, Chapter 3) shows this is possible, Morgan (1935) describes another such construction.

Theorem 6.15
There exists a 1-set in $\mathbb{R}^2$ (with positive $\mathscr{H}^1$-measure) with projection on every line of $\mathscr{L}^1$-measure zero.

Proof. Let $k \ge 2$ be an integer, and let r be the smallest integer such that $1/k > 2\pi/r$. Let ϕ be the angle $2\pi/r$. Suppose that P is a parallelogram with side lengths a and $a/2k$ and internal angles $\frac{1}{2}(\pi \pm \phi)$. We describe a sequence of operations that replace P by subsets made up of many smaller parallelograms similar to P.

First inscribe $2k$ parallelograms inside P, each similar to P with shorter sides of lengths $a/(2k)^2$ equally spaced on the longer sides of P and with longer sides, of lengths $a/2k$, making angles $\frac{1}{2}(\pi - \phi)$ with the longer sides of P. Do exactly the same thing with each of these new parallelograms to get

Fig. 6.1

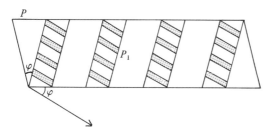

$(2k)^2$ parallelograms, each similar to P but oriented at angle ϕ to P and with side lengths $a/(2k)^2$ and $a/(2k)^3$, see Figure 6.1. Let P_1 denote the union of these parallelograms and let T denote the operation on the parallelogram P that replaces P by the subset P_1. We note that the sum of the lengths of the longer sides of the parallelograms forming P_1 is a, and that

$$\mathcal{L}^1(\mathrm{proj}_\theta P_1) \le a/k \quad \text{if } 0 \le \theta \le \phi,$$

where the angle θ is measured from the direction of the base line of P.

We now perform operation T on each of the parallelograms forming P_1 to produce a subset P_2 made up of $(2k)^4$ parallelograms, each similar to P and oriented at angle 2ϕ to P, and with longer sides of lengths $a/(2k)^4$. By construction, $\mathcal{L}^1(\mathrm{proj}_\theta P_2) \le a/k$ if $\phi \le \theta \le 2\phi$, so as $P_2 \subset P_1$ this is true for $0 \le \theta \le 2\phi$.

We continue in this way, performing operation T on the parallelograms comprising P_{j-1} to get P_j, a union of parallelograms, each oriented at angle $j\phi$ to P, for $j = 2, \ldots, r$. Since $\phi = 2\pi/r$, the $(2k)^{2r}$ parallelograms making up P_r are homothetic (similar and similarly situated) to P. The base lengths of the parallelograms sum to a and, further, $\mathcal{L}^1(\mathrm{proj}_\theta P_r) \le a/k$ for $0 \le \theta \le 2\pi$. Denote by T_k the operation on P that replaces P by the subset P_r constructed in this way.

We next define an associated operation U_k which replaces a line segment S_0 of length a by a large collection of parallel line segments. As before let r be the smallest integer such that $1/k > 2\pi/r = \phi$, and erect a parallelogram P with sides a and $a/2k$ and internal angles $\frac{1}{2}(\pi \pm \phi)$ on the segment S_0 as base. Perform the operation T_k on P to get a set $P_r = Q$, say, and let S be the set of base lines of the parallelograms that form Q. From the corresponding properties of the parallelograms it follows that S consists of $(2k)^{2r}$ line segments of total length a, and that S projects to a set of measure, at most, a/k in all direction. Let U_k be the operation that replaces the segment S_0 by the set of segments S.

Finally, let $\{k_j\}$ be a sequence of integers tending to infinity. Take a unit segment of the plane and perform operation U_{k_1} on it to get a set S_1

consisting of line segments, with Q_1 as the set formed by the corresponding parallelograms. Continue in this way to obtain line segment sets S_j and corresponding parallelogram sets Q_j, where S_j is the aggregate of the line segments obtained by performing the operation U_{k_j} on each line segment in S_{j-1}.

Let $E = \bigcap_{j=1}^{\infty} Q_j$. Then E has the required properties: certainly $\mathscr{L}^1(\text{proj}_\theta E) = 0$ for all θ, since $\mathscr{L}^1(\text{proj}_\theta Q_j) \leq 1/k_j$ for all θ. To see that the Borel set E is a 1-set, we first note that since E may be covered by m parallelograms, each of diameter, at most, $2/m$ for arbitrarily large values of m, $\mathscr{H}^1(E) \leq 2$. If $\{V_i\}$ is a cover of E by open convex sets, then, by a compactness argument, $Q_j \subset \bigcup_i V_i$ for some j. By construction, the sum of the base lengths of the parallelograms of Q_j that intersect V_i can be, at most, $3|V_i|$, so $\frac{1}{3} \leq \sum |V_i|$. It follows that $\frac{1}{3} \leq \mathscr{H}^1(E)$ and E is a 1-set. (In fact $\mathscr{H}^1(E) = 2^{-\frac{1}{2}}$.) $\square$

Gillis (1936*b*) shows that the set described in Theorem 6.15 also projects to a set of (angular) measure zero from every point in the plane.

We have already seen an irregular 1-set that projects to the unit interval in at least two directions (Theorem 3.32). Marstrand (1954*a*) showed that an irregular 1-set could have $\mathscr{L}^1(\text{proj}_\theta E) > 0$ for a set of θ of dimension 1, improving an earlier result of Gillis (1934*a*). Such a set will be described in Theorem 8.17.

Theorems 6.16 and 6.17 below are the natural generalizations of Theorems 6.10 and 6.13 to higher dimensions:

Theorem 6.16
Let E be a regular k-set in $\mathbb{R}^n$, where k is an integer. Then $\mathscr{L}^k(\text{proj}_\Pi E) > 0$ for almost all $\Pi \in G_{n,k}$.

Theorem 6.17
Let E be an irregular k-set in $\mathbb{R}^n$, where k is an integer. Then $\mathscr{L}^k(\text{proj}_\Pi E) = 0$ for almost all $\Pi \in G_{n,k}$.

These two theorems are proved (for general Hausdorff measures) by Federer (1947) in his mammoth paper. Much of the work on the structure of s-sets in n-dimensions depends on these projection results, and a simpler proof would be a very useful addition to the literature.

6.5 Further variants
We describe briefly some variations on the ideas that we have described in this chapter.

First, recall the idea of projective transformation. Given any line L in the plane, we may find a projective bijection ψ_L of the plane extended by the line at infinity that maps the points at infinity onto L. The mapping ψ_L transforms lines to lines, and, except at points mapped onto the line at infinity, is locally analytic. Thus if proj_x denotes projection from the point x (i.e. the mapping that takes y to the point of intersection of the half-line from x through y with the unit circle centre x), we may transform Theorems 6.8, 6.10 and 6.13 under ψ_L to get results on projections from the points of the line L. For example, transforming Theorem 6.8(a) we see that if E is a plane Souslin set with $\dim E = s \leq 1$, then $\dim(\text{proj}_x E) = s$ (as a subset of the unit circle) for almost all $x \in L$ for each line L, and thus (taking a complete set of parallel lines) for almost all $x \in \mathbb{R}^2$.

Marstrand (1954a, Section 6) investigated the intersection of a set with the lines through its points in a way that exhibits the local structure of s-sets. He proved that if E is a plane s-set with $s > 1$, then almost every line through almost every point of E intersects E in a set of dimension $s - 1$ and finite $\mathscr{H}^{s-1}$-measure. Mattila (1975b) generalized this to higher dimensions in a natural way and in a later paper (1981) couched some of these results in terms of capacities.

One may also consider the intersection of s-sets with curves rather than straight lines. For example, if f is a Lipschitz function so that $f^{-1}(x)$ is a curve or surface, it is possible to obtain inequalities of the form

$$\int \mathscr{H}^s(E \cap f^{-1}(x)) \mathrm{d}\mathscr{H}^t(x) \leq c\mathscr{H}^{s+t}(E)$$

see Federer (1969), and also Mattila (1984a, 1984b) for capacity analogues. If Γ is a curve and E is a plane s-set with $s > 1$, then $\dim(E \cap \sigma(\Gamma)) = s - 1$ for a set of rigid motions σ of positive measure. Mattila (1984a) also describes higher-dimensional analogues and shows how, at least in dimension, such results fail for intersections with congruent copies of *irregular* sets. On the other hand, if E itself is rectifiable, we are led to the results of classical integral geometry described in Federer (1969) or Santaló (1976). Mattila (1982) also studies these ideas when E is a self-similar set (see Section 8.3). In this very special case the dimension of intersection of E with curves or surfaces behaves with surprising consistency.

Exercises on Chapter 6

6.1 Prove that capacities are subadditive, that is, $C_t(E \cup E') \leq C_t(E) + C_t(E')$ for $E, E' \subset \mathbb{R}^n$. Give an example to show that equality need not hold even if E and E' are disjoint Borel sets, so that C_t is not a Borel measure.

6.2 Let Γ be a rectifiable curve in $\mathbb{R}^2$. Let $n(t, \theta)$ denote the number of intersections of Γ with the line in direction θ at distance t from the origin. Obtain Poincare's formula of integral geometry, that

$$L(\Gamma) = \tfrac{1}{2} \int_{\theta=0}^{\pi} \int_{t=-\infty}^{\infty} n(t,\theta) \mathrm{d}t \mathrm{d}\theta.$$

6.3   Divide the unit square into k^2 smaller squares of side $1/k(k \geq 3)$ in the obvious way. Shade one square in each column of the array so that the centres of the squares are not collinear. Construct a set E by a Cantor-type process by repeatedly replacing each shaded square by a similar copy of the whole figure and taking the limit set. Show that E is an *irregular* 1-set.

7
Besicovitch and Kakeya sets

7.1 Introduction

The Kakeya problem has an interesting history. In 1917 Besicovitch was working on problems on Riemann integration, and was confronted with the following question: if f is a Riemann integrable function defined on the plane, is it always possible to find a pair of orthogonal coordinate axes with respect to which $\int f(x, y)\mathrm{d}x$ exists as a Riemann integral for all y, and with the resulting function of y also Riemann integrable? Besicovitch noticed that if he could construct a compact set F of plane Lebesgue measure zero containing a line segment in every direction, this would lead to a counter-example: For assume, by translating F if necessary, that F contains no segment parallel to and rational distance from either of a fixed pair of axes. Let f be the characteristic function of the set F_0 consisting of those points of F with at least one rational coordinate. As F contains a segment in every direction on which both F_0 and its complement are dense, there is a segment in each direction on which f is not Riemann integrable. On the other hand, the set of points of discontinuity of F is of plane measure zero, so f is Riemann integrable over the plane by the well-known criterion of Lebesgue.

Besicovitch (1919) succeeded in constructing a set, known as a 'Besicovitch set', with the required properties. Owing to the unstable situation in Russia at the time, his paper received limited circulation, and the construction was later republished in *Mathematische Zeitschrift* (1928).

At about the same time, Kakeya (1917) and Fujiwara & Kakeya (1917) mentioned the problem of finding the area of the smallest convex set inside which a unit segment can be reversed, that is, manoeuvred to lie in its original position but rotated through 180° without leaving the set. They conjectured that the equilateral triangle of unit height was the smallest such set, and recorded an observation of Kubota, that if the convexity condition was dropped, then a smaller set was possible, namely a three-cusped hypercycloid. The conjecture in the convex case was proved by Pál (1921) who reiterated the more interesting question without the convexity assumption. This became known as the Kakeya problem.

Shortly after Besicovitch's departure from Russia in 1924, it was realized that a simple modification to the Besicovitch set yielded a solution to the

Kakeya problem of arbitrarily small measure (Besicovitch (1928) and Perron (1928)) and the problem was solved in an unexpected manner.

Perhaps the most significant subsequent development was again due to Besicovitch (1964*a*) who found a fundamental relationship between Besicovitch sets and the geometric measure theory described so far in this book. Using the techniques of polar reciprocity he demonstrated that the existence of Besicovitch sets follows as a simple dual result to the projection theorems for irregular linearly measurable sets. It seems surprising that this dual relationship had not been noticed before by Besicovitch (who had been interested in both subjects for years) or by other researchers.

Numerous papers have appeared describing variants on these problems. For example, plane sets of measure zero containing copies of all polygons or circles of all radii have been constructed, and the possibility of higher-dimensional analogues has been considered. Recently, such ideas have been used to answer hitherto difficult questions of harmonic analysis.

It is worth mentioning that in 1958 the Kakeya problem was selected by the Mathematical Association of America to be the subject of the first of their films. Any opportunity of seeing this charming film, narrated by Besicovitch himself, should not be missed.

7.2 Construction of Besicovitch and Kakeya sets

This section describes the construction of a Besicovitch set, that is, a set of measure zero containing a line segment in every direction. Besicovitch's original construction (1919, 1928) has been simplified considerably (Perron (1928), van Alphén (1942), Rademacher (1962), Schoenberg (1962*b*), Besicovitch (1963*c*), Cunningham (1971) and Fisher (1973)). The basic idea behind all the constructions is to form a 'Perron tree', a figure obtained by splitting an equilateral triangle of unit height into many smaller triangles of the same height by dividing up the base, and then sliding these elementary triangles varying distances along the base line L (see Figure 7.2). Such a figure certainly contains line segments of unit length in all directions at angles at least $60°$ to L. The difficulty is to show that a set of arbitrarily small measure may be obtained in this way.

Lemma 7.1

Let T_1 and T_2 be adjacent triangles with bases on a line L, with base lengths b and heights h. Take $\frac{1}{2} < \alpha < 1$. Then if T_2 is slid a distance $2(1 - \alpha)b$ along L to overlap T_1, the resulting figure S consists of a triangle T homothetic (that is, similar and similarly situated) to triangle $T_1 \cup T_2$ with $\mathscr{L}^2(T) = \alpha^2 \mathscr{L}^2(T_1 \cup T_2)$, and two auxiliary triangles. The reduction in area effected by *replacing* $T_1 \cup T_2$ *by* S *is given by*

$$\mathcal{L}^2(T_1 \cup T_2) - \mathcal{L}^2(S) = \mathcal{L}^2(T_1 \cup T_2)(1 - \alpha)(3\alpha - 1).$$

($\mathcal{L}^2$ *is plane area or Lebesgue measure.*)

Proof. The situation is illustrated in Figure 7.1. The difference in area between the two figures is very easily found using elementary geometry. (Note that the calculation is affine invariant.) □

Fig. 7.1

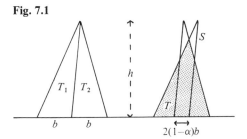

Theorem 7.2

Let T be a triangle with base on a line L. Divide the base of T into 2^k equal segments and join each point of division to the opposite vertex to form 2^k elementary triangles $T_1, \ldots, T_{2^k}$. By choosing k large enough it is possible to translate the elementary triangles along L to positions such that the area of the resulting (closed) figure S is as small as desired. Further, if V is an open set containing T, this may be achieved with $S \subset V$.

Proof. This construction involves repeated applications of the previous Lemma for a fixed value of α to be specified later. We first work with consecutive pairs of elementary triangles. For each i ($1 \le i \le 2^{k-1}$) move T_{2i} along L relative to T_{2i-1} to get a figure S_i^1 consisting of a triangle T_i^1 (homothetic to $T_{2i-1} \cup T_{2i}$) and two auxiliary triangles. By Lemma 7.1 this may be done so that $\mathcal{L}^2(T_i^1) = \alpha^2 \mathcal{L}^2(T_{2i-1} \cup T_{2i})$ and with a reduction in area (comparing S_i^1 with $T_{2i-1} \cup T_{2i}$) of $(1 - \alpha)(3\alpha - 1) \mathcal{L}^2(T_{2i-1} \cup T_{2i})$. For the second stage of the construction we work with consecutive S_i^1. For $1 \le i \le 2^{k-2}$ translate S_{2i}^1 relative to S_{2i-1}^1 to get S_i^2. Since one side of T_{2i-1}^1 is parallel and equal to the opposite side of T_{2i}^1 we may, by Lemma 7.1, do this so that S_i^2 includes a triangle T_i^2 with $\mathcal{L}^2(T_i^2) = \alpha^2[\mathcal{L}^2(T_{2i-1}^1) + \mathcal{L}^2(T_{2i}^1)]$ and so that the overlap of T_{2i-1}^1 and T_{2i}^1 results in a reduction in area of at least

$$(1 - \alpha)(3\alpha - 1)[\mathcal{L}^2(T_{2i-1}^1) + \mathcal{L}^2(T_{2i}^1)]$$
$$= (1 - \alpha)(3\alpha - 1)\alpha^2 \mathcal{L}^2(T_{4i-3} \cup T_{4i-2} \cup T_{4i-1} \cup T_{4i}).$$

Continue in this way; at the $(r + 1)$th stage we obtain S_i^{r+1} by moving S_{2i}^r relative to S_{2i-1}^r $(1 \leq i \leq 2^{k-r})$ so that the overlap of triangles T_{2i-1}^r and T_{2i}^r results in an area reduction of at least $(1 - \alpha)(3\alpha - 1)\alpha^{2r}$ times the total area of the elementary triangles moved to form S_{2i-1}^r and S_{2i}^r. We end up with a single figure S_1^k which we take for S. The stages of construction in the case of $k = 3$ are shown in Figure 7.2.

Comparing the areas of $\bigcup_i S_i^r$ for successive values of r, we see that

$$\mathscr{L}^2(S) \leq \mathscr{L}^2(T) - (1 - \alpha)(3\alpha - 1)(1 + \alpha^2 + \cdots + \alpha^{2(k-1)})\mathscr{L}^2(T)$$

$$= \left(1 - \frac{(3\alpha - 1)(1 - \alpha^{2k})}{(1 + \alpha)}\right)\mathscr{L}^2(T).$$

By choosing α close enough to 1 (to make $(3\alpha - 1)/(1 + \alpha)$ close to 1) and then taking k large enough, we can make $\mathscr{L}^2(S)$ as small as required.

Finally, if the base length of T is b, then fixing T_1 and locating the other elementary triangles relative to T_1, no triangle will have been moved by a distance of more than b during the construction. Thus by dividing the original triangle into subtriangles of base lengths, at most, ε, and performing the operation described above on each of these subtriangles, we may obtain a figure of arbitrarily small area by moving none of the elementary triangles more than ε. If ε is chosen small enough, this figure will be contained in the open set V. $\square$

Fig. 7.2

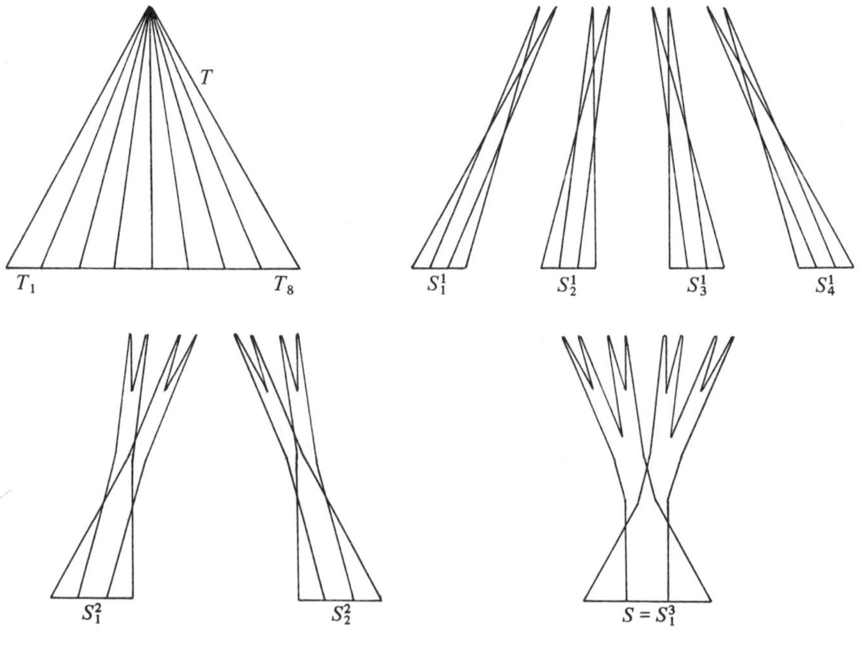

An interesting problem is to determine the smallest area, A_m, that can be obtained by sliding triangles along the base line when T is divided into m elementary triangles. Schoenberg (1962a, 1962b) has shown that $A_m = O(1/\log m)$, but it is not even known if A_m decreases monotonically in m. However, all we need for the above proof to work is that $\lim_{m \to 0} A_m = 0$. We use Theorem 7.2 to construct a Besicovitch set.

Theorem 7.3

There exists a plane set of Lebesgue measure zero which contains a unit segment in every direction.

Proof. We construct a set F of measure zero containing unit segments in all directions in a $60°$ sector. Taking the union of F with congruent copies of F rotated through $60°$ and $120°$ then gives a set with the required properties.

The construction consists of repeated applications of Theorem 7.2. Let S_1 be an equilateral triangle of unit height based on a line L. Let V_1 be an open set containing S_1 such that $\mathscr{L}^2(\overline{V}_1) \leq 2\mathscr{L}^2(S_1)$ (where the bar denotes closure). By Theorem 7.2 we may divide up S_1 into a number of elementary triangles of unit height and with bases on L, and translate the elementary triangles along L to form a closed figure S_2 contained in V_1 with $\mathscr{L}^2(S_2) \leq 2^{-2}$. Since S_2 is a finite union of triangles, we may find an open set V_2 such that $S_2 \subset V_2 \subset V_1$ and $\mathscr{L}^2(\overline{V}_2) \leq 2\mathscr{L}^2(S_2)$. Similarly, we may split up each elementary triangle of S_2 into further elementary triangles which may be moved along L in such a way that the resulting figure S_3 is contained in V_2 with $\mathscr{L}^2(S_3) \leq 2^{-3}$. We enclose S_3 in an open set V_3 with $\mathscr{L}^2(\overline{V}_3) \leq 2\mathscr{L}^2(S_3)$, and continue in the same way. We obtain a sequence of figures $\{S_i\}$, each a finite union of elementary triangles based on L and of unit height, and a sequence of open sets $\{V_i\}$ such that for each i,

$$S_i \subset V_i \subset V_{i-1}$$

and

$$\mathscr{L}^2(\overline{V}_i) \leq 2\mathscr{L}^2(S_i) \leq 2^{-i+1}.$$

Let F be the closed set $F = \bigcap_{i=1}^{\infty} \overline{V}_i$. We claim that F has the desired properties. Certainly, $\mathscr{L}^2(F) = 0$. By construction, each S_i, and thus each $\overline{V}_i$, contains a unit segment in any direction making an angle of $60°$ or more with L. We must show that this is also true of F. Let θ be some such direction, and for each i let M_i be a unit segment in direction θ with $M_i \subset \overline{V}_i$. By a standard compactness argument we may assume, taking a subsequence if necessary, that $\{M_i\}$ converges in the obvious sense to M, a unit segment in direction θ. Since $\{\overline{V}_i\}$ is a decreasing sequence, $M_i \subset \overline{V}_j$ if $i \geq j$, so as $\overline{V}_j$ is closed, $M \subset \overline{V}_j$ for each j. Thus $M \subset \bigcap_{i=1}^{\infty} \overline{V}_i = F$, as required. $\square$

Various other methods of constructing Besicovitch sets have been found: Kahane (1969) noticed that such a set could be obtained by joining the points of a Cantor set in the *x*-axis to the points of a parallel Cantor-like set, see also Kinney (1970) and Alexander (1975). A further method, using duality, will be described in Section 7.3.

The same sort of construction may be used to form a *Nikodym set*; that is, a subset F of the unit square with $\mathscr{L}^2(F) = 1$ such that for each $x \in F$ there is a line through x intersecting F in the single point x (we say each point of F is 'linearly accessible'). Nikodym's (1927) complicated construction, which answered a question of Banach, was simplified by Davies (1952*a*) who showed that it was even possible to find such a set with uncountably many such lines through each point of the set. (These constructions are also described in de Guzmán (1975, 1979, 1981), Casas (1978) and Casas & de Guzmán (1981).) Davies (1952*a*) uses similar ideas to obtain the following surprising result, quoted here without proof.

Theorem 7.4
Given any plane set G of finite $\mathscr{L}^2$-measure, there exists a set F consisting of a union of straight lines such that $G \subset F$ and $\mathscr{L}^2(G) = \mathscr{L}^2(F)$.

Larman (1971*a*) also uses such techniques in the construction of his 'impossible set'. He exhibits a compact set F made up of a disjoint union of closed line segments in $\mathbb{R}^n (n \geq 3)$ such that $\mathscr{L}^n(F) > 0$ but $\mathscr{L}^n(G) = 0$, where G is the union of the corresponding open segments. Thus the measure of F is concentrated in the endpoints of the segments. (Such constructions are impossible in $\mathbb{R}^2$.)

Next, we use Theorem 7.2 to solve the Kakeya problem. We require the following easy lemma.

Lemma 7.5
Let L_1 and L_2 be parallel lines in the plane. Then, given $\varepsilon > 0$, there is a set E containing L_1 and L_2 with $\mathscr{L}^2(E) < \varepsilon$ such that a unit segment may be moved continuously from L_1 to L_2 without leaving E.

Proof. Let x_1 and x_2 be points on L_1 and L_2. Let E be the set consisting of L_1, L_2, the segment M joining x_1 and x_2 and the unit sectors centred at x_i lying between L_i and M ($i = 1, 2$), see Figure 7.3. It is very easy to see that the total area of E can be made as small as desired by taking x_1 and x_2 sufficiently far apart. Further, a unit segment may be moved from L_1 to L_2 by a rotation in the first sector, a translation along M, and a rotation in the second sector. □

Fig. 7.3

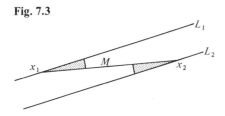

Theorem 7.6
Given $\varepsilon > 0$, there exists a set E with $\mathscr{L}^2(E) < \varepsilon$ inside which a unit segment may be moved continuously to lie in its original position but rotated through $180°$.

Proof. We construct a set E_0 with $\mathscr{L}^2(E_0) < \frac{1}{3}\varepsilon$ inside which it is possible to manoeuvre a unit segment to a position at an angle of $60°$ with its original direction. Taking three copies of this construction we can then obtain a set E as required. (We may need Lemma 7.5 if translations are needed between the three component sets.)

If T is an equilateral triangle of unit height based on the line L, we may divide T into $m = 2^k$ elementary triangles and slide them along L to positions $T'_1, \ldots, T'_m$ in such a way that $\mathscr{L}^2(\bigcup_1^m T'_i) < \varepsilon/6$ (see Theorem 7.2). Let $E_1 = \bigcup_1^m T'_i$. Note that for each i, one side of T'_i is parallel to the opposite side of T'_{i+1}. Therefore, by Lemma 7.5 we may, for each i, add a set of measure, at most $\varepsilon/6m$ to E_1 to allow a unit segment to be moved from T'_i to T'_{i+1}. This gives a set E_0 of measure, at most, $\frac{1}{6}\varepsilon + \frac{1}{6}(m-1)\varepsilon/m < \frac{1}{2}\varepsilon$ inside which a unit segment may be rotated through $60°$, as required. $\qquad\square$

7.3 The dual approach

This section relates Besicovitch sets to the theory of linearly measurable sets described in Chapters 3 and 6. The basic idea is to parametrize lines by points in such a way that the projection of a set of points E in some direction is geometrically similar to the intersection of the lines parametrized by E with some fixed line. Then we may examine the set formed as the union of the lines (the 'line set') by projecting E in various directions.

Besicovitch (1964a) used the technique of polar reciprocity in this way. Let $L(x)$ be the line at distance $1/|x|$ from the origin and perpendicular to the radius vector x; thus $L(x)$ is the polar line to x with respect to the unit circle C. Points at infinity represent lines through the origin in the natural way. The fundamental property of polar reciprocity is that $x \in L(y)$ if and only if $y \in L(x)$. It is easy to see that if proj_θ denotes projection onto the line L_θ

through the origin in direction θ, then, for any set E, the intersection of the line set $\{L(x):x\in E\}$ with L_θ is simply the geometric inverse of the set $\mathrm{proj}_\theta E$ with respect to C. Hence the projection theorems of Chapter 6 may be used to study intersection properties of sets of lines.

For technical reasons we introduce an alternative form of duality, though the fundamental ideas are the same. For the time being we work in the plane, with (x, y), etc. denoting Cartesian coordinates.

If $(a, b)\in\mathbb{R}^2$ let $L(a, b)$ denote the set of points on the line $y = a + bx$. If E is a subset of $\mathbb{R}^2$ let $L(E)$ be the line set $\bigcup_{(a,b)\in E}L(a, b)$. If c is a constant and L_c is the line $x = c$, then

$$L(a, b)\cap L_c = (c, a + bc) = (c, (a, b)\cdot(1, c)),$$

where '$\cdot$' denotes the usual scalar product in $\mathbb{R}^2$. Thus if $E\subset\mathbb{R}^2$,

$$L(E)\cap L_c = \{(c, (a, b)\cdot(1, c)):(a, b)\in E\},$$

so that the set $L(E)\cap L_c$ is geometrically similar to $\mathrm{proj}_\theta E$ with a ratio of similitude of $(1 + c^2)^{\frac{1}{2}}$, where $c = \tan\theta$. (As usual, proj_θ denotes projection onto the line L_θ through the origin at an angle θ to the x-axis.) In particular,

$$\mathcal{L}^1(L(E)\cap L_c) = 0 \Leftrightarrow \mathcal{L}^1(\mathrm{proj}_\theta E) = 0. \tag{7.1}$$

Further, considering projections onto the y-axis, we see that if $(a, b)\in\mathbb{R}^2$, then $\mathrm{proj}_{\pi/2}(a, b) = b$ is simply the gradient of the line $L(a, b)$. Thus if E is any set of points and $b\in\mathrm{proj}_{\pi/2}E$, then $L(E)$ contains a line of gradient b.

The following theorem gives the fundamental relationship between linearly measurable sets and line sets.

Theorem 7.7
Let E be a 1-set in $\mathbb{R}^2$. Then $L(E)$ is an $\mathcal{L}^2$-measurable subset of $\mathbb{R}^2$. If E is irregular then $\mathcal{L}^2(L(E)) = 0$, and if E is regular then $\mathcal{L}^2(L(E)) > 0$.

Proof. The only slight awkwardness is checking the measurability of $L(E)$, though, as usual, this presents no serious difficulty (see the comment of Croft (1965) in his review of Besicovitch's paper).

It is easy to see that $L(E)$ is open if E is an open set and is closed if E is closed. Hence if E is a G_δ-set (a countable intersection of open sets) or an F_σ-set (a countable union of closed sets), then $L(E)$ is of the same type.

Suppose $\mathcal{H}^1(E) = 0$. By the regularity of $\mathcal{H}^1$ (Theorem 1.6) we may find a G_δ-set E_0 with $E\subset E_0$ and $\mathcal{H}^1(E_0) = 0$. By Lemma 6.1 $\mathcal{L}^1(\mathrm{proj}_\theta(E_0)) = 0$ for all θ, so by (7.1), $\mathcal{L}^1(L(E_0)\cap L_c) = 0$ for all c. But $L(E_0)$ is a G_δ-set and so is plane measurable. Fubini's theorem is valid so $\mathcal{L}^2(L(E_0)) = 0$, giving that $L(E)\subset L(E_0)$ is $\mathcal{L}^2$-measurable with $\mathcal{L}^2(L(E)) = 0$.

Now let E be any 1-set. By regularity, $E = E_0\cup F$, where E_0 is an F_σ-set and $\mathcal{H}^1(F) = 0$. Thus $L(E) = L(E_0) \cup L(F)$ is $\mathcal{L}^2$-measurable, as the union of two measurable sets.

If E is an irregular 1-set, then $\mathcal{L}^1(\text{proj}_\theta E) = 0$ for almost all θ, by Theorem 6.13. By the duality principle (7.1), $\mathcal{L}^1(L(E) \cap L_c) = 0$ for almost all c. As $L(E)$ is $\mathcal{L}^2$-measurable, Fubini's theorem implies that $\mathcal{L}^2(L(E)) = 0$.

If E is a regular 1-set, then by Theorem 6.10, $\mathcal{L}^1(\text{proj}_\theta E) > 0$ for almost all θ, so $\mathcal{L}^1(L(E) \cap L_c) > 0$ for almost all c, by (7.1). In this case the measurability of $L(E)$, together with Fubini's theorem, gives that $\mathcal{L}^2(L(E)) > 0$. $\quad\square$

Besicovitch (1964a) further showed that if E is a regular 1-set, then $L(E)$ has infinite plane measure, provided the multiplicity of the covering by the line set is taken into account. However, Davies (1965) was able to construct a regular 1-set E with $\mathcal{L}^2(L(E)) < \infty$, showing that the multiplicity assumption is crucial. (Although these papers represent lines by polar reciprocity, the different parametrizations are equivalent under a locally real-analytic bijection, so that the regularity of the sets is independent of the parametrization used.)

We now give a simple construction of a Besicovitch set using duality. This strengthens Theorem 7.3, since we obtain a set of measure zero containing a complete line (rather than just a segment) in every direction.

Theorem 7.8
There exists a subset of the plane of Lebesgue measure zero containing a line in every direction.

Proof. Let E be an irregular 1-set in $\mathbb{R}^2$ such that the projection of E onto the y-axis, $\text{proj}_{\pi/2} E$, contains the segment $-1 \le y \le 1$. For example, E might be the set of Theorem 3.32, together with its reflection in the x-axis. Consider the line set $L(E)$; by Theorem 7.7 $\mathcal{L}^2(L(E)) = 0$. On the other hand, the remarks prior to Theorem 7.7 imply that $L(E)$ contains lines of all gradients between -1 and 1. Taking the union of $L(E)$ with a congruent copy rotated through $\frac{1}{2}\pi$ we get a subset of $\mathbb{R}^2$ with the required properties. $\quad\square$

Our next application of duality, due to Davies (1971), shows that Besicovitch sets are necessarily quite large.

Theorem 7.9
Let F be a subset of the plane containing a line in every direction. Then F has Hausdorff dimension 2.

Proof. Every set is contained in a G_δ-set of the same dimension, so we lose no generality in assuming F to be G_δ. Thus the set $E = \{(a, b) : L(a, b) \subset F\}$ $= \bigcap_{r=1}^{\infty} \{(a, b) : L(a, b) \cap B_r(0) \subset F \cap B_r(0)\}$ is also G_δ and so is Borel measurable. Since F contains lines in every direction, $\text{proj}_{\pi/2} E$ is the entire y-axis and $\mathcal{H}^1(E) = \infty$ by Lemma 6.1. By the projection theorem, Theorem 6.8(a), it follows that $\dim(\text{proj}_\theta E) = 1$ for almost all θ, so, by the duality principle, $\dim(L(E) \cap L_c) = 1$ for almost all c. By Theorem 5.8, $\dim L(E) = 2$, so as $L(E) \subset F$, $\dim F = 2$. $\quad\square$

A slight modification of the above proof shows that if F contains a segment in every direction, then $\dim F = 2$ (see Davies (1971)). In fact, an even stronger result is true, namely that if $\dim F < 2$ then F intersects all lines in almost all directions in a set of 1-dimensional measure zero. This can be shown using the same basic idea, but employing Marstrand's variation of the projection theorem mentioned at the end of Section 6.3. (This technique is used in higher dimensions by Falconer (1980*b*).)

Another curve-packing problem was considered by Besicovitch & Rado (1968) (in Besicovitch's last paper) and by Kinney (1968), who gave direct constructions of plane sets of measure zero containing circles of every radius. It was Davies (1972) who realized that this problem lends itself to the dual approach.

Theorem 7.10
There exists a subset of the plane of zero Lebesgue measure which contains a circle of every radius.

Proof. Let E be a bounded irregular 1-set, such as that described in Theorem 3.32, with projection onto the x-axis, $\text{proj}_\theta E$, containing the interval $0 \leq x \leq 1$. Define a mapping $\psi : \mathbb{R}^2 \to \mathbb{R}^2$ by

$$\psi(x, y) = (x(1 + y^2)^{\frac{1}{2}}, y).$$

For each x with $0 \leq x \leq 1$ there exists y such that $(x, y) \in E$. Thus for any such value of x we may find a point of the form $(x(1 + y^2)^{\frac{1}{2}}, y)$ in $\psi(E)$, that is, a point $(a, b) \in \psi(E)$ such that $a^2 - x^2 b^2 = x^2$ or $x = |a|(1 + b^2)^{-\frac{1}{2}}$. But this is precisely the perpendicular distance of the line $L(a, b)$ from the origin, so we conclude that the line set $L(\psi(E))$ contains lines at all distances between 0 and 1 from the origin. Further, since ψ is a real-analytic mapping, $\psi(E)$ is an irregular 1-set (ψ maps rectifiable curves to rectifiable curves), so, by Theorem 7.7, $\mathcal{L}^2(L(\psi(E))) = 0$. Now let F be the set obtained by inverting $L(\psi(E))$ with respect to the origin (in other words, F is the image of $L(\psi(E))$ under the transformation given in polar coordinates by $(r, \theta) \to (1/r, \theta)$). Inversion maps sets of measure zero to sets of measure zero, and also transforms straight lines at distance x from the origin to circles of radius

$1/(2x)$ through the origin. Thus F has measure zero and contains circles of all radii greater than $\frac{1}{2}$. By taking a countable union of sets similar to F with similarity ratios tending to zero, we obtain a set with the required properties. □

In the light of this construction it is natural to ask whether or not it is possible for a plane set of measure zero to contain circles with centres at every point. This interesting and difficult problem is as yet unsolved.

Davies's theorem, Theorem 7.4, has a surprising dual. Essentially, we may find a set that has any desired projection or 'shadow' in each direction. This generalizes a result of Talagrand (1980).

Theorem 7.11

Let G_θ be a subset of $\mathbb{R}$ for each $\theta \in [0, \pi)$ and suppose that the set $\{(\theta, y) : y \in G_\theta\}$ is plane Lebesgue measurable. Then there exists a set $E \subset \mathbb{R}^2$ such that for almost all θ we have $\{u\boldsymbol{\theta} : u \in G_\theta\} \subset \mathrm{proj}_\theta E$ and $\mathscr{L}^1(G_\theta) = \mathscr{L}^1(\mathrm{proj}_\theta E)$, where $\boldsymbol{\theta}$ is the unit vector in direction θ.

Proof. Let

$$G = \{(c, y) : (1 + c^2)^{-\frac{1}{2}} y \in G_\theta, \text{ where } c = \tan\theta\}. \tag{7.2}$$

By Theorem 7.4 we may find a line set, $L(E)$, say, containing G and with $\mathscr{L}^2(G) = \mathscr{L}^2(L(E))$. As G is plane measurable, it follows from Fubini's theorem that $\mathscr{L}^1(G \cap L_c) = \mathscr{L}^1(L(E) \cap L_c)$ for almost all c, with $G \cap L_c \subset L(E) \cap L_c$. By the duality principle, the set $L(E) \cap L_c$ is similar to $\mathrm{proj}_\theta E$, where $c = \tan\theta$, with a ratio of similitude of $(1 + c^2)^{\frac{1}{2}}$. Thus by (7.2) $\mathrm{proj}_\theta E$ and G_θ differ by measure zero for almost all θ. □

Finally, we use duality to investigate the possibility of higher-dimensional analogues of Besicovitch sets.

Theorem 7.12

A subset of $\mathbb{R}^3$ of $\mathscr{L}^3$-measure zero cannot contain a translate of every plane.

Proof. After suitable parametrization of the planes in $\mathbb{R}^3$, the proof is very similar to that of Theorem 7.9. For $(a, b, c) \in \mathbb{R}^3$ let $\Pi(a, b, c)$ denote the set of points on the plane $z = a + bx + cy$, and if $E \subset \mathbb{R}^3$ let $\Pi(E) = \bigcup_{(a,b,c) \in E} \Pi(a, b, c)$. If d, e are constants and $L_{d,e}$ is the line $x = d, y = e$, then

$$\Pi(a, b, c) \cap L_{d,e} = (d, e, a + bd + ce) = (d, e, (a, b, c) \cdot (1, d, e)),$$

where '·' is the scalar product in $\mathbb{R}^3$. Hence, if $E \subset \mathbb{R}^3$,

$$\Pi(E) \cap L_{d,e} = \{(d, e, (a, b, c) \cdot (1, d, e)) : (a, b, c) \in E\},$$

so that $\Pi(E) \cap L_{d,e}$ is geometrically similar to $\text{proj}_\theta E$, the projection of E onto the line in the direction $\boldsymbol{\theta}$ of $(1, d, e)$. Thus

$$\mathcal{L}^1(\Pi(E) \cap L_{d,e}) = 0 \Leftrightarrow \mathcal{L}^1(\text{proj}_\theta E) = 0. \qquad (7.3)$$

Now suppose that the set $F \subset \mathbb{R}^3$ contains a translate of every plane. Without loss of generality we may assume that F is a G_δ-set, in which case it is easy to see that the set $E = \{(a, b, c) : \Pi(a, b, c) \subset F\}$ is a Borel set. If F contains planes perpendicular to all unit vectors and Π is the plane $x = 0$, then $\text{proj}_\Pi E$ is the whole of Π, so, by Lemma 6.1, $\mathcal{H}^2(E) = \infty$. By Theorem 6.9(b), $\mathcal{L}^1(\text{proj}_\theta E) > 0$ for almost all $\boldsymbol{\theta}$, and so $\mathcal{L}^1(\Pi(E) \cap L_{d,e}) > 0$ for almost all $(d, e) \in \mathbb{R}^2$, by (7.3). Just as in Theorem 7.9, the measurability of E implies that $\Pi(E)$ is $\mathcal{L}^3$-measurable, so, by Fubini's theorem, $\mathcal{L}^3(\Pi(E)) > 0$. As $\Pi(E) \subset F$, the set F must have positive measure. $\qquad \square$

It is not hard to modify the above proof to show that any subset of $\mathbb{R}^3$ of measure zero intersects all planes perpendicular to almost all directions (in the sense of spherical measure) in sets of plane measure zero.

The existence of higher-dimensional analogues of Besicovitch sets is a question of considerable interest. We call a subset of $\mathbb{R}^n$ an (n, k)-Besicovitch set if it is of n-dimensional Lebesgue measure zero, but nevertheless contains a translate of every k-dimensional subspace of $\mathbb{R}^n$. We showed in Theorem 7.8 that $(2, 1)$-Besicovitch sets exist, and for any n the Cartesian product of such a set with $\mathbb{R}^{n-2}$ is an $(n, 1)$-Besicovitch set. Marstrand (1979b) proved that $(3, 2)$-sets could not exist by a direct covering method and, simultaneously, Falconer (1980a) showed, using Fourier transforms, that there are no (n, k)-Besicovitch sets if $k > \frac{1}{2}n$. Oberlin & Stein (1982) also mention the $(n, n-1)$ case in connection with functional analysis. Theorem 7.12 generalizes easily enough to show that $(n, n-1)$-Besicovitch sets cannot exist, but the use of duality in the general (n, k) case is more awkward. Falconer (1980b) proved that (n, k)-Besicovitch sets cannot exist if $2 \leq k \leq n-1$ by using a form of duality, together with the projection theorems, with the projections factorized by a common intermediate space. We state here the most general form of this result:

Theorem 7.13
Suppose $2 \leq k \leq n-1$, and let F be a subset of $\mathbb{R}^n$ of n-dimensional Lebesgue measure zero. Then, for almost all k-dimensional subspaces Π of $\mathbb{R}^n$ (in the sense of the usual invariant measure on the Grassmann manifold $G_{n,k}$), every translate of Π intersects F in a set of k-dimensional measure zero.

7.4 Generalizations

This section surveys some of the many generalizations and variations that have been inspired by the Besicovitch and Kakeya constructions. (Some of this material is also discussed in Croft (unpublished mineographed notes), Davies (1971), Cunningham (1974) and Taylor (1975).)

A general form of problem is: given a collection of geometric figures, is it possible to find a plane set of measure zero containing a translate, or even just a congruent copy, of every figure in the collection? The Besicovitch set provides an answer in the case of translates of all straight lines.

Kinney (1968) observed that, since the Cantor set E is of Lebesgue measure zero and contains all distances in the range $(0,1)$, the 'Cantor tartan' $(E \times [0,1]) \cap ([0,1] \times E)$ is of plane measure zero and contains a congruent copy of every rectangle with sides of less than unit length. Taking a countable union of similar sets gives a set of measure zero containing copies of all rectangles. Ward (1970a, 1970b) constructed sets of measure zero containing congruent copies of all polygons; indeed, he produced a set of Hausdorff dimension $2 - 1/n$ that contained a congruent copy of every n-sided polygon. Davies (1971) then modified Besicovitch's original construction to give a set of measure zero containing a translate of every polygonal arc, necessarily of dimension 2, by Theorem 7.9. Marstrand (1972) proved the strongest results of this type in a very general paper applicable to any topological measure space. He showed, among other things, that there exists a set of measure zero containing a translate of every countable union of straight lines, and also that there is a set of Hausdorff dimension as small as 1 containing a congruent copy of every such union.

Another natural problem to consider is that of packing copies of *curved* arcs into sets of measure zero. We have already seen (Theorem 7.10 and associated remarks) that there exist sets of measure zero containing circles of all radii, and we have mentioned the unsolved problem for circles of all centres. Other obvious questions are whether a set of measure zero can contain copies of all ellipses, conics, or plane curves of degree k for each $k > 1$. Whilst it seems to be generally believed that such sets cannot exist, these problems are of considerable difficulty, not least because of the non-linearities involved, and nothing has yet been proved. Some relatively weak results on the dimensions of such sets and on packing higher-dimensional surfaces in $\mathbb{R}^n$ for $n \geq 3$ are given by Falconer (1982). Other such packing results, usually in three or more dimensions, can be deduced from maximal theorems in harmonic analysis, see Section 7.5.

Undoubtedly the most significant development in this area is due to Marstrand (1979a) who solved the 'worm' problem ('what are the minimal comfortable living quarters for a unit worm?') by showing that for any subset F of $\mathbb{R}^n$ of n-dimensional measure zero ($n \geq 2$) there is some smooth (C^∞) curve of unit length which has no congruent copy contained in F. Indeed, there is such a curve that cannot be transformed into a subset of F

by any invertible real-analytic mapping defined on a domain containing the curve. The complicated proof uses the idea of the entropy of a totally bounded metric space, that is, an estimate of the number of sets of diameter ε required to cover the space for small ε.

Brief mention should also be made of the opposite type of packing problem: do there exist sets of positive measure containing no congruent or similar copy of certain specified sets? The best-known result of this type, due to Steinhaus (1920), is that any subset of the line of positive measure contains all distances in the range $(0, c)$ for some positive c. (See Exercise 1.7; Besicovitch & Miller (1948), Besicovitch (1948), Eggleston (1949) and Besicovitch & Taylor (1952) give various generalizations.) This property is shared by the Cantor set of measure zero. The Steinhaus result follows from the Lebesgue density theorem, Theorem 1.13, which also implies that any subset of the line of positive length contains a similar copy of any finite set of points. Erdös has posed the intriguing question of whether this remains true for a countable (convergent) sequence of points and Falconer (1984) gives a partial solution. Several other variations on this theme are examined by Davies, Marstrand & Taylor (1960) and Darst & Goffman (1970).

The problem of finding the smallest *convex* set containing copies of all figures of certain types really belongs to convexity rather than to geometric measure theory. We merely point out that in the paper by Pál (1921), which stimulated much of the research described in this chapter, it was shown that the equilateral triangle was the smallest plane convex set containing a unit segment in every direction; the analogous problem in higher dimensions remains unsolved. Related 'convexity' problems are considered by Eggleston (1957), Besicovitch (1965b) and Wetzel (1973), among others.

We return to the Kakeya problem of manoeuvring a unit segment through 180° inside a set that is as small as possible. The ultimate solution was provided by Cunningham (1971) who constructed a simply connected Kakeya set of arbitrarily small measure contained in the unit disc. The same paper considers the minimal *starshaped* Kakeya set, and shows that the lower bound for its area lies between $\pi/108$ and $(5 - 2\sqrt{2})\pi/24$, superseding earlier estimates of Walker (1952), Blank (1963) and Cunningham & Schoenberg (1965). (A set is starshaped if it has a point joined to all other points of the set by line segments contained in the set.) The minimal convex Kakeya set is again the equilateral triangle of unit height (Pál (1921)), with the 3-dimensional analogue unknown.

Cunningham (1974) considers a trio of problems worthy of mention. First he finds the area of the smallest subset of a spherical surface inside which an arc of a great circle can be reversed. The minimal area (previously established as zero for arcs smaller than a semicircle by Wilker (1971)) depends on the angle subtended by the arc. The second problem discussed in the paper involves finding small (plane) sets inside which a circular arc of fixed angle but variable radius can be contracted to a point. In Cunningham's final variation a 'bird' is defined as a central line segment (the 'body') with a further segment (the 'wings') attached non-rigidly at each end. It is shown that it is possible to move the bird continuously across any bounded set E in such a way that the body passes over all of E whilst the wings remain in a set of arbitrarily small measure. This problem has applications in harmonic analysis, and is closely related to the work of Davies (1952a) on linearly accessible sets.

7.5 Relationship with harmonic analysis

This chapter would not be complete without some mention of its relationship to harmonic analysis. On the one hand it is sometimes possible to obtain packing results from more general norm bounds for certain integrals. On the other, the Kakeya construction has been used to provide counter-examples to some major conjectures in harmonic analysis. We use the phrase 'harmonic analysis' loosely, to describe a vast area of mathematics including Fourier analysis and the theories of differentiation, maximal operators and multipliers. We can do no more than give a small number of examples to illustrate the relationship.

In our first example we deduce Theorem 7.12, on the non-existence of higher-dimensional Besicovitch sets from a functional-analytic inequality. Suppose $f(x)$ is a measurable function defined on $\mathbb{R}^3$. Write $\|f\|_p = [\int_{\mathbb{R}^3} |f|^p dx]^{1/p}$ in the usual way, where $1 \le p < \infty$. Let $F(t, \theta)$ denote the integral of F over the plane perpendicular to the unit vector θ and distance t from the origin; $F(t, \theta)$ exists for almost all t for all θ by a simple application of Fubini's theorem. It is not hard to show, by integrating the Fourier transform of F with respect to t over all unit vectors θ and using the fact that the transform of an integrable function is bounded, that

$$\int_\theta \operatorname*{ess\,sup}_t |F(t, \theta)| d\theta \le c(\|f\|_1 + \|f\|_2) \tag{7.4}$$

for some constant c independent of F. (See Falconer (1980a) or Oberlin & Stein (1982).)

Now suppose E is a set of measure zero, and write $F_0(\theta)$ for the supremum of the plane outer measure of the intersection of E with the planes perpendicular to θ. Routine methods show that F_0 is measurable with respect to spherical measure. By the regularity of Lebesgue measure we may find an open set V containing E and of measure less than ε, thus if f is the characteristic function of V, then $\|f\|_1 = \|f\|_2 < \varepsilon$. As V is open,

$F_0(\theta) \leq \text{ess sup } F(t, \theta)$, so, by (7.4),
$$\int F_0(\theta) d\theta \leq 2\varepsilon c.$$

Since ε is arbitrary and $F_0(\theta)$ is non-negative, we conclude that $F_0(\theta) = 0$ for almost all θ, in other words all planes in almost all directions intersect E in sets of plane measure zero.

This sort of argument may be used the other way round. We know (Theorem 7.3) that there exists a plane set of measure zero containing a line in every direction, so if an estimate such as (7.4) were to hold for functions on $\mathbb{R}^2$ (with θ now denoting a unit vector in the plane), a similar argument to the above would lead to a contradiction. Hence the Besicovitch set provides a counter-example to the plausible conjecture that (7.4) is valid for plane functions.

As stated in Theorem 7.13 and the preceding remarks, (n, k)-Besicovitch sets do not exist if $k > 1$. It is of interest that the method outlined above may be adapted to show this if $k > \frac{1}{2}n$, but not otherwise (see Falconer (1980a)). This suggests, perhaps, that if $2 \leq k \leq \frac{1}{2}n$ the non-existence of (n, k)-Besicovitch sets is an intrinsically geometric property, rather than a consequence of a more general functional-analytic result.

We move on to spherical averages and questions of packing spheres. If f is a measurable function on $\mathbb{R}^n$ let $F_r(x)$ denote the average of f over the spherical surface of centre x and radius r, that is, the integral of f over this surface divided by the $(n-1)$-dimensional surface area of the sphere. Let

$$F(x) = \sup_{r > 0} F_r(x). \tag{7.5}$$

Stein (1976) proved that

$$\|F\|_p \leq c \|f\|_p, \tag{7.6}$$

provided that $n/(n-1) < p$ and $n \geq 3$. (If, instead, we had defined $F_r(x)$ to be the average of f over the *ball* of centre x and radius r, then $F(x)$ would be the Hardy–Littlewood maximal operator, and (7.6) the classical maximal inequality, holding for all n and $1 < p < \infty$, see de Guzmán (1975).)

We use exactly the same argument as above, but starting with (7.6) rather than (7.4), to deduce that if $n \geq 3$ and E is any subset of $\mathbb{R}^n$ of zero n-dimensional Lebesgue measure, then, for almost all $x \in \mathbb{R}^n$, all spheres with x as centre intersect E in sets of $(n-1)$-dimensional surface measure zero.

If (7.6) were proved for $n = 2$ and some value of p this would imply that a plane set of measure zero could not contain a circle centred at every point, solving the problem mentioned earlier. On the other hand, if such a packing of circles of null measure were constructed, this would imply that (7.6) fails to hold for all p rather than just for $p \leq 2$, which is all that is known at present in $\mathbb{R}^2$.

The proof of (7.6) depends crucially on the non-vanishing of the curvature of the sphere. Similar results hold for averages over other 'well-curved' surfaces (see Stein & Wainger (1978)), with consequent packing results, but fail, for example, for the surface of a cube.

Many generalizations of Lebesgue's density theorem concern averages over sets other than balls. One plausible conjecture (certainly true for continuous functions) might be that if f is a 'reasonable' function on the plane, then

$$\inf \left\{ \frac{1}{\mathscr{L}^2(R)} \int_R f \, d\mathscr{L}^2 : R \text{ is a rectangle centred at} \right.$$

$$\left. x \quad \text{with } 0 < |R| \leq r \right\} \tag{7.7}$$

converges to $f(x)$ as r tends to 0 for almost all x, with a similar result for the supremum. However, a consequence of the existence of Nikodym sets (discussed after Theorem 7.3) is that such results are not even true for characteristic function of sets. (We say that the class of all rectangles does not form a differentiation basis.) For let E be a Nikodym set, that is, a subset of the unit square of measure 1 with each of its points on some line intersecting E in a single point. By regularity we may find a closed subset F of E with the same accessibility property and with $\mathscr{L}^2(F) > 0$. Let f be the characteristic function of F. As the complement of F is open, it is easy to see that, by taking thin rectangles with major axes along the 'exceptional lines', the infimum (7.7) tends to zero as r tends to 0 for all x in F.

The Besicovitch set has provided an important and surprising counter-example in the theory of Fourier multipliers. Let T be the linear transformation on the space of pth power integrable functions on $\mathbb{R}^2$ defined in Fourier transform notation by

$$(Tf)\hat{\;}(d) = \chi_B(d)\hat{f}(d),$$

where χ_B is the characteristic function of the unit disc. It was known that

$$\| Tf \|_p \leq c \| f \|_p \tag{7.8}$$

for $p = 2$, and was widely expected that such an inequality would also hold for all $\frac{4}{3} < p < 4$. However, Fefferman (1971) used the Besicovitch set in an ingenious way to demonstrate that (7.8) fails if $p \neq 2$. The interested reader is referred to Fefferman's clearly written paper for further details.

A similar sort of procedure was followed by Mitjagin & Nikisin (1973) to obtain, for $p < 2$, a pth power integrable function f on the unit square with its partial Fourier sums unbounded almost everywhere, that is, with

$$\overline{\lim_{r \to \infty}} \left| \sum_{k^2 + j^2 < r} a_{kj} \exp\left(2\pi i(xk + yj)\right) \right| = \infty$$

for almost all $(x, y) \in \mathbb{R}^2$, where $\{a_{kj}\}$ are the 2-dimensional Fourier coefficients of f. This is in sharp contrast to the 1-dimensional case where the Carleson–Hunt theorem states that if f is pth power integrable for some p with $1 < p < \infty$, then the Fourier series of f converges pointwise almost everywhere.

Finally, we mention an application to functional analysis: T is a hypernormal operator on a Hilbert space (that is, with $T^*T - TT^* \geq 0$ and no non-trivial reducing space of T normal), if and only if the spectrum of T has positive measure in all neighbourhoods of all its points. Putnam (1974) thus uses the Nikodym set to prove a positive result rather than just to provide a counter-example.

To explore the relevant areas of harmonic analysis more deeply, the reader is referred to the survey article by Stein & Wainger (1978), the American Mathematical Society conference proceedings on harmonic analysis, edited by Wainger & Weiss (1979), and the books by de Guzmán (1975, 1981). It seems certain that geometric measure theory and harmonic analysis will continue to influence each other greatly.

Exercises on Chapter 7

7.1   Show that there exists a subset E of $\mathbb{R}^n (n \geq 2)$ inside which a unit segment may be manoeuvred to lie in any direction, with $\mathscr{L}^n(E)$ arbitrarily small.

7.2   Let S be a Borel subset of $[0, \pi)$. Let F be a subset of $\mathbb{R}^2$ containing a line in every direction in S. Use duality to show that dim $F \geq 1 + \dim S$.

7.3   Show that there exists a subset of $\mathbb{R}^2$ of $\mathscr{L}^2$-measure zero that contains a different straight line passing through every point on the x-axis.

7.4   Show that if E is a Borel subset of $\mathbb{R}^2$ with dim $E > 1$, then the set of polar lines of the points of E (with respect to any fixed circle) cover a set of infinite plane Lebesgue measure.

7.5   If f is a plane measurable function let $F_r(x)$ denote the average of f over the circumference of the circle of centre x and radius r. It may be shown (Wainger (1979)) that if E is the Cantor set,

$$\left\| \sup_{r \in E} F_{1+r}(x) \right\|_2 \leq c \|f(x)\|_2$$

for some constant c. Deduce that it is not possible for a plane set of Lebesgue measure zero to contain a circle centred at each point of $\mathbb{R}^2$ with radius $1 + r$ for some $r \in E$.

8

Miscellaneous examples of fractal sets

8.1 Introduction

This chapter surveys examples of sets of fractional dimension which result from particular constructions or occur in other branches of mathematics or physics and relates them to earlier parts of the book. The topics have been chosen very much at the author's whim rather than because they represent the most important occurrences of fractal sets. In each section selected results of interest are proved and others are cited. It is hoped that this approach will encourage the reader to follow up some of these topics in greater depth elsewhere. Most of the examples come from areas of mathematics which have a vast literature; therefore in this chapter references are given only to the principal sources and to recent papers and books which contain further surveys and references.

8.2 Curves of fractional dimension

In this section we work in the (x, y)-coordinate plane and investigate the Hausdorff dimension of Γ, the set of points $(x, f(x))$ forming the graph of a function f defined, say, on the unit interval.

If f is a function of bounded variation, that is, if $\sum_{i=1}^{m} |f(x_i) - f(x_{i-1})|$ is bounded for all dissections $0 = x_0 < x_1 < \cdots < x_m = 1$, then we are effectively back in the situation of Section 3.2; Γ is a rectifiable curve and so a regular 1-set. However, if f is a sufficiently irregular, though continuous, function it is possible for Γ to have dimension greater than 1. In such cases it can be hard to calculate the Hausdorff dimension and measure of Γ from a knowledge of f. However, if f satisfies a Lipschitz condition it is easy to obtain an upper bound.

Theorem 8.1

Suppose that

$$|f(x+h) - f(x)| \leq ch^{2-s} \tag{8.1}$$

for all x and all h with $0 < h \leq h_0$, where c and h_0 are positive constants. Then $\mathscr{H}^s(\Gamma) < \infty$.

Proof. Let I be any interval on the x-axis of length $h < h_0$. It follows from (8.1), by taking a column of squares above I, that the part of Γ given by

$\{(x, f(x)) : x \in I\}$ may be covered by, at most, $h^{-1}ch^{2-s} + 1$ squares of side h. Thus, dividing $[0, 1]$ into m equal parts of length $h = 1/m < h_0$, we see that

$$\mathcal{H}^s_{2^{\frac{1}{2}}h}(\Gamma) \leq m(h^{-1}ch^{2-s} + 1)(2^{\frac{1}{2}}h)^s \leq c2^{1+s/2}$$

for a sequence of values of h tending to 0. Thus $\mathcal{H}^s(\Gamma) \leq c2^{1+s/2} < \infty$.   □

The easiest way to obtain a function whose graph has a fine structure is to add together a sequence of functions which oscillate increasingly rapidly. Thus if $\sum_1^\infty |a_i| < \infty$ and $\lambda_i \to \infty$, the function defined by the trigonometric series

$$f(x) = \sum_{i=1}^\infty a_i \sin(\lambda_i x) \tag{8.2}$$

might be expected to have a graph of dimension greater than 1 if the a_i and λ_i are chosen suitably. Perhaps the best-known example of this type is the function

$$f(x) = \sum_{i=1}^\infty \lambda^{(s-2)i} \sin(\lambda^i x), \tag{8.3}$$

where $1 < s < 2$ and $\lambda > 1$, constructed by Weierstrass to be continuous but nowhere differentiable. It seems likely that the Weierstrass function has graph of dimension s, but this does not appear to have been proved rigorously. (The dimension cannot exceed s, see Exercise 8.3.) A variant of the Weierstrass function

$$f(x) = \sum_{i=-\infty}^\infty \lambda^{(s-2)i}(1 - \cos \lambda^i x)$$

was introduced by Mandelbrot (1977). This function has the scaling property that $f(\lambda x) = \lambda^{2-s} f(x)$ for all x; again the dimension of the graph ought to be s. Berry & Lewis (1980) give computer realizations of such functions and cite references to various physical applications.

For ease of calculation it is convenient to replace the sine functions in (8.2) by periodic functions of a slightly different form. Let g be the 'zig-zag' function of period 4 defined on $\mathbb{R}$ by

$$g(4k + x) = \begin{cases} x & (0 \leq x < 1) \\ 2 - x & (1 \leq x < 3) \\ x - 4 & (3 \leq x < 4), \end{cases}$$

where k is an integer and $0 \leq x < 4$. (Note in particular that g has derivative equal to 1 in modulus at all non-integral x.) Then we study the functions

$$f(x) = \sum_{i=1}^\infty a_i g(\lambda_i x) \tag{8.4}$$

instead of (8.2). Besicovitch & Ursell (1937) have established the Hausdorff dimension of the graphs of certain functions of this type:

Theorem 8.2
Let Γ be the graph of the function

$$f(x) = \sum_{i=1}^{\infty} \lambda_i^{s-2} g(\lambda_i x) \tag{8.5}$$

for $x \in [0, 1]$ where $1 < s < 2$. Suppose that $\{\lambda_i\}$ is a sequence of positive numbers with λ_{i+1}/λ_i increasing to infinity and $\log \lambda_{i+1}/\log \lambda_i \to 1$. Then $\dim \Gamma = s$.

Proof. If

$$\lambda_{k+1}^{-1} \leq h < \lambda_k^{-1}, \tag{8.6}$$

then

$$|f(x+h) - f(x)| \leq \sum_{1}^{k} \lambda_i^{s-2} |g(\lambda_i(x+h)) - g(\lambda_i x)| + 2 \sum_{k+1}^{\infty} \lambda_i^{s-2}$$

$$\leq h \sum_{1}^{k} \lambda_i^{s-1} + 2 \sum_{k+1}^{\infty} \lambda_i^{s-2}$$

$$\leq 2h\lambda_k^{s-1} + 4\lambda_{k+1}^{s-2},$$

provided that k is large enough. Thus, by (8.6),

$$|f(x+h) - f(x)| \leq 2h^{2-s} + 4h^{2-s} = 6h^{2-s}$$

if h is sufficiently small. We conclude from Theorem 8.1 that $\mathcal{H}^s(\Gamma) < \infty$.

The lower estimate of the dimension of Γ is rather more awkward to obtain. Let S be a square with sides of length h and parallel to the coordinate axes. Let I be the interval of projection of S onto the x-axis. We show that the Lebesgue measure of the set $E = \{x : (x, f(x)) \in S\}$ cannot be too big.

Define the partial sums

$$f_k(x) = \sum_{1}^{k} \lambda_i^{s-2} g(\lambda_i x).$$

We assume throughout that we are dealing with values of k large enough to ensure that $\lambda_{k+1} \geq 2\lambda_k \geq 2$, that

$$|f(x) - f_k(x)| = \left| \sum_{k+1}^{\infty} \lambda_i^{s-2} g(\lambda_i x) \right| \leq \sum_{k+1}^{\infty} \lambda_i^{s-2} \leq 2\lambda_{k+1}^{s-2}, \tag{8.7}$$

and that (for non-exceptional x)

$$|f_k'(x)| = \left| \sum_{1}^{k} \lambda_i^{s-1} g'(\lambda_i x) \right| \geq \lambda_k^{s-1} - \sum_{1}^{k-1} \lambda_i^{s-1} \geq \tfrac{1}{2}\lambda_k^{s-1}. \tag{8.8}$$

First suppose that the square S has side $h = \lambda_k^{-1}$ for some such k. Let m be the integer such that

$$\lambda_{k+m}^{s-2} \leq h = \lambda_k^{-1} < \lambda_{k+m-1}^{s-2}. \tag{8.9}$$

Certainly, $m \geq 1$. On the other hand, since λ_{i+1}/λ_i is increasing,

$$\left(\frac{\lambda_{k+1}}{\lambda_k}\right)^{(m-1)(2-s)} \lambda_k^{2-s} < \left(\frac{\lambda_{k+m-1}}{\lambda_{k+m-2}} \cdots \frac{\lambda_{k+1}}{\lambda_k} \lambda_k\right)^{2-s} = \lambda_{k+m-1}^{2-s} < \lambda_k,$$

so that

$$\left(\frac{\lambda_{k+1}}{\lambda_k}\right)^{(m-1)(2-s)} < \lambda_k^{s-1} < \left(\frac{\lambda_k}{\lambda_{k-1}} \cdots \frac{\lambda_2}{\lambda_1} \lambda_1\right)^{s-1} < \left(\frac{\lambda_{k+1}}{\lambda_k}\right)^{(k-1)(s-1)} \lambda_1^{s-1}.$$

Hence, on taking logarithms,

$$m \leq ak, \tag{8.10}$$

where a is independent of k.

If $m = 1$, then by (8.7) $(x, f(x))$ can lie in S only if $(x, f_k(x))$ lies in the rectangle S_1 obtained by extending S a distance $2\lambda_{k+1}^{s-2} \leq 2h$ above and below. The derivative $f_k'(x)$ changes sign, at most, once in the interval I. On each section on which $f_k'(x)$ is of constant sign $|f_k'(x)| \geq \frac{1}{2}\lambda_k^{s-1}$ so $(x, f_k(x))$ can lie in S_1 for x in an interval of length, at most, $2\lambda_k^{1-s}$ times the height of S_1. Thus

$$\mathscr{L}^1(E) \leq 2 \cdot 2\lambda_k^{1-s} \cdot 5h = 20\,h^s. \tag{8.11}$$

If $m > 1$ we can divide I into, at most, two parts, on each of which $f_k'(x)$ is of constant sign. The height of S_1 is $h + 4\lambda_{k+1}^{s-2} \leq 5\lambda_{k+1}^{s-2}$ by (8.9), so in each part we need only consider x belonging to a subinterval of length $2\lambda_k^{1-s} \cdot 5\lambda_{k+1}^{s-2}$ when seeking points of E. We divide each of these subintervals further into parts on which $f_{k+1}'(x)$ is also of constant sign. In this way we obtain, at most, $2\lambda_k^{1-s} \cdot 5\lambda_{k+1}^{s-2} \cdot \frac{1}{2}\lambda_{k+1} + 1 \leq 6(\lambda_{k+1}/\lambda_k)^{s-1}$ new intervals from each of the old.

If $m > 2$ we repeat the process to obtain from each of the last set of intervals, at most, $6(\lambda_{k+2}/\lambda_{k+1})^{s-1}$ intervals on each of which $f_{k+2}'(x)$ is also of constant sign. Proceeding in this way we eventually see that E is covered by, at most,

$$2 \cdot 6^{m-1}\left(\frac{\lambda_{k+1}}{\lambda_k} \cdot \frac{\lambda_{k+2}}{\lambda_{k+1}} \cdots \frac{\lambda_{k+m-1}}{\lambda_{k+m-2}}\right)^{s-1} = 2 \cdot 6^{m-1}\left(\frac{\lambda_{k+m-1}}{\lambda_k}\right)^{s-1}$$

intervals on each of which $f_{k+m-1}'(x)$ is of constant sign. By (8.7) it follows that on each such interval, $(x, f(x)) \in S$ only if $(x, f_{k+m-1}(x)) \in S_2$, where S_2 is the rectangle formed by extending S a distance $2\lambda_{k+m}^{s-2}$ above and below. The height of S_2 is $h + 4\lambda_{k+m}^{s-2} \leq 5h$, so, by considering the gradient of f_{k+m-1}

on each such interval and using (8.8) and (8.9), we have

$$\mathscr{L}^1(E) \leq 2 \cdot 6^{m-1} \left(\frac{\lambda_{k+m-1}}{\lambda_k} \right)^{s-1} \cdot 5h \cdot 2\lambda_{k+m-1}^{1-s}$$

$$\leq 20 \cdot 6^{m-1} h^s \leq 20 \cdot 6^{ak} h^s.$$

Thus there exist constants b and c such that $\mathscr{L}^1(E) \leq cb^k h^s$ if $h = \lambda_k^{-1}$.

Now suppose that S is a square of side h where $\lambda_{k+1}^{-1} < h \leq \lambda_k^{-1}$. It follows from above that if $t < s$,

$$\mathscr{L}^1(E) \leq cb^k \lambda_k^{-s} = c\lambda_{k+1}^{-t} \frac{\lambda_{k+1}^t}{\lambda_k^{(s+t)/2}} \frac{b^k}{\lambda_k^{(s-t)/2}}.$$

Hence

$$\mathscr{L}^1(E) \leq c_1 h^t, \tag{8.12}$$

since $\lambda_k^{(s+t)/2}$ increases faster than λ_{k+1}^t and $\lambda_k^{(s-t)/2}$ increases faster than b^k for large k, in view of the stated growth conditions on λ_k.

If $\{U_i\}$ is any cover of Γ enclose each U_i in a square S_i of side equal to $|U_i|$. Writing $E_i = \{x : (x, f(x)) \in S_i\}$, we must have that $[0, 1] \subset \bigcup_i E_i$. Thus

$$\Sigma |U_i|^t = \Sigma 2^{-\frac{1}{2}t} |S_i|^t \geq c_1^{-1} \Sigma \mathscr{L}^1(E_i) \geq c_1^{-1},$$

by (8.12). Hence $\mathscr{H}^t(\Gamma) > c_1^{-1} > 0$ if $t < s$, and we conclude that $\dim \Gamma = s$. $\square$

Note that an identical argument shows that, with the same conditions on the λ_i, the dimension of the graph of

$$f(x) = \sum_{i=1}^{\infty} \lambda_i^{s-2} g(\lambda_i x + \theta_i)$$

is s, where the θ_i are fixed 'phases'.

Besicovitch & Ursell (1937) use a very similar method to show that if in (8.5) $\lambda_{i+1} = \lambda_i^\mu$ for all i, where $\mu = (s-1)(2-t)/(2-s)(t-1)$ where $1 < t < s$, then $\dim \Gamma = t$. Some further computations are given by Love & Young (1937) and Kline (1945).

Various other definitions of the 'dimension' of curves have been introduced and compared with Hausdorff dimension. For example, one might consider the lower or upper limit of $\sum_{i=1}^{m} |f(x_i) - f(x_{i-1})|^s$ as the maximum interval length of the dissection tends to zero (see Ville (1936)). Alternatively, if $\Gamma(x, x+h)$ denotes the part of Γ corresponding to the interval $[x, x+h]$, the limits of $\mathscr{H}^s(\Gamma(x, x+h))/h^s$ as $h \to 0$ have some interesting properties (Besicovitch (1964b, 1965a, 1967)).

8.3 Self-similar sets

Many of the classical fractal sets are 'self-similar', built up of pieces geometrically similar to the entire set but on a smaller scale. The simplest example is the Cantor set (see Section 1.4). Here $E \cap [0, \frac{1}{3}]$ and $E \cap [\frac{2}{3}, 1]$ are similar to E but scaled by a factor $\frac{1}{3}$; $E \cap [0, \frac{1}{9}], E \cap [\frac{2}{9}, \frac{1}{3}], E \cap [\frac{2}{3}, \frac{7}{9}]$ and $E \cap [\frac{8}{9}, 1]$ are similar to E with a scale factor $\frac{1}{9}$, and so on. Two examples in the plane are depicted in Figures 8.1 and 8.2; Figure 8.1 shows the familiar Koch or snowflake curve. Computer realizations of many other beautiful self-similar sets are pictured in the books by Mandelbrot (1977, 1982) and in the paper by Dekking (1982), and Stepney (1984) gives a computer program for drawing such sets.

Examples of self-similar sets have been known throughout the century, but only recently have attempts been made to put their theory on a systematic basis. Dekking (1982) describes a general method of doing this using endomorphisms of words in free groups. Here, however, we give a

Fig. 8.1

(a)

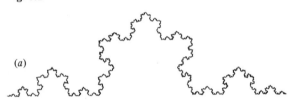

(b)

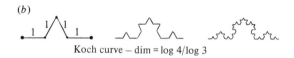

Koch curve – dim = log 4/log 3

Fig. 8.2

(a)

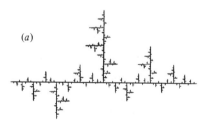

(b)

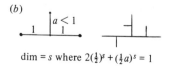

dim = s where $2(\frac{1}{2})^s + (\frac{1}{2}a)^s = 1$

version of Hutchinson's (1981) elegant treatment (see also Moran (1946) and Marion (1979)).

A mapping $\psi : \mathbb{R}^n \to \mathbb{R}^n$ is called a *contraction* if $|\psi(x) - \psi(y)| \le c|x - y|$ for all $x, y \in \mathbb{R}^n$, where $c < 1$. Clearly, any contraction is a continuous mapping. We call the infimum value of c for which this inequality holds for all x, y the *ratio* of the contraction. A contraction that transforms every subset of $\mathbb{R}^n$ to a geometrically similar set is called a *similitude*. Thus a similitude is a composition of a dilation, a rotation, a translation and perhaps a reflection; the ratio is then simply the scale factor of the similitude.

For the purposes of this section we call a set $E \subset \mathbb{R}^n$ *invariant* for a set of contractions $\psi_1, \dots, \psi_m$ if $E = \bigcup_1^m \psi_j(E)$. If in addition the contractions are similitudes and for some s we have $\mathcal{H}^s(E) > 0$ but $\mathcal{H}^s(\psi_i(E) \cap \psi_j(E)) = 0$ for $i \ne j$, then E is *self-similar*. (This measure condition ensures that the self-similar features of E are not lost by overlaps.) We show that for any finite set of contractions there exists a unique non-empty compact invariant set E, and that for a set of similitudes satisfying an 'open set condition' the set E is a self-similar s-set for a value of s calculable from the contraction ratios. To fix ideas, the Cantor set is the unique compact set invariant under the similitudes of the real line

$$\psi_1(x) = x/3, \quad \psi_2(x) = (2 + x)/3.$$

If $\{\psi_j\}_1^m$ is a set of contractions, let ψ denote the transformation of subsets of $\mathbb{R}^n$ defined by

$$\psi(F) = \bigcup_1^m \psi_j(F).$$

We denote the iterates of ψ by $\psi^0(F) = F$ and $\psi^{k+1}(F) = \psi(\psi^k(F))$ for $k \ge 0$.

Note that the work of this section is equally valid for contractions defined on a compact subset of $\mathbb{R}^n$.

Theorem 8.3
Given a set of contractions $\{\psi_j\}_1^m$ on $\mathbb{R}^n$ with contraction ratios $r_j < 1$ there exists a unique non-empty compact set E such that

$$E = \psi(E) = \bigcup_1^m \psi_j(E). \tag{8.13}$$

Further, if F is any non-empty compact subset of $\mathbb{R}^n$ the iterates $\psi^k(F)$ converge to E in the Hausdorff metric as $k \to \infty$.

Proof. Let $\mathscr{C}$ be the class of all non-empty compact subsets of $\mathbb{R}^n$. By the completeness section of the proof of the Blaschke selection theorem, Theorem 3.16, $\mathscr{C}$ becomes a complete metric space when endowed with the

Hausdorff metric δ. If $F_1, F_2 \in \mathscr{C}$, then from the definition of δ,

$$\delta(\psi(F_1), \psi(F_2)) = \delta\left(\bigcup_1^m \psi_j(F_1), \bigcup_1^m \psi_j(F_2)\right)$$

$$\leq \max_j \delta(\psi_j(F_1), \psi_j(F_2))$$

$$\leq (\max_j r_j)\delta(F_1, F_2).$$

(As the ψ_j are contractions it follows that if F_1 is contained in the δ-parallel body of F_2, then $\psi_j(F_1)$ is contained in the $r_j\delta$-parallel body of $\psi_j(F_2)$.) Since $\max_j r_j < 1$, ψ is a contraction mapping on $\mathscr{C}$. By the contraction mapping theorem for complete metric spaces there is a unique $E \in \mathscr{C}$ with $\psi(E) = E$ and, moreover, $\delta(\psi^k(F), E) \to 0$ as $k \to \infty$ for any $F \in \mathscr{C}$. $\square$

It may be shown that the invariant set E is the closure of the set of fixed points of the mappings $\psi_{j_1} \circ \psi_{j_2} \circ \ldots \circ \psi_{j_k}$ taken for all finite sequences $\{j_1, j_2, \ldots, j_k\}$ with $1 \leq j_i \leq m$. (We adopt the convention that $(\psi_1 \circ \psi_2)(x) = \psi_1(\psi_2(x))$, etc.)

Suppose now that s is the number such that

$$\sum_1^m r_j^s = 1.$$

Raising this to the kth power we obtain the useful identity

$$\sum_{j_1 \ldots j_k} (r_{j_1} r_{j_2} \ldots r_{j_k})^s = 1, \tag{8.14}$$

where the sum is over all k-tuples $\{j_1 \ldots j_k\}$ with $1 \leq j_i \leq m$. For any such sequence and any set F write

$$F_{j_1 \ldots j_k} = \psi_{j_1} \circ \ldots \circ \psi_{j_k}(F).$$

The following lemma may be thought of as an analogue of Theorem 8.3 for measures, indeed Hutchinson (1981) gives a proof using the contraction mapping theorem.

Lemma 8.4

There exists a Borel measure μ with support contained in E such that $\mu(\mathbb{R}^n) = 1$ and such that for any measurable set F,

$$\mu(F) = \sum_{j=1}^m r_j^s \mu(\psi_j^{-1}(F)). \tag{8.15}$$

Proof. Choose $x \in E$ and write $x_{j_1 \ldots j_k} = \psi_{j_1} \circ \ldots \circ \psi_{j_k}(x)$. For $k = 1, 2, \ldots$

define a positive linear functional on the space $\mathscr{F}$ of continuous functions on E by

$$\phi_k(f) = \sum_{j_1\ldots j_k} (r_{j_1}\ldots r_{j_k})^s f(x_{j_1\ldots j_k}). \tag{8.16}$$

If $f \in \mathscr{F}$, then f is uniformly continuous on E, so we may, given $\varepsilon > 0$, find p such that f varies by less than ε over each $E_{j_1\ldots j_k}$ whenever $k \geq p$ (recall that $|E_{j_1\ldots j_k}| \leq (\max_j r_j)^k |E|)$. Since $x_{j_1\ldots j_k} \in E_{j_1\ldots j_p}$ if $k \geq p$ it follows from (8.16) and (8.14) that $|\phi_k(f) - \phi_{k'}(f)| \leq \varepsilon$ if $k, k' \geq p$. By the general principle of convergence $\{\phi_k(f)\}_k$ is convergent for each f and the limit defines a positive linear functional on $\mathscr{F}$. By the Riesz representation theorem, Theorem 6.2, there exists a Borel measure μ such that

$$\int f\,\mathrm{d}\mu = \lim_{k\to\infty} \phi_k(f) \tag{8.17}$$

for $f \in \mathscr{F}$. Putting $f \equiv 1$ it is clear from (8.16) that $\mu(\mathbb{R}^n) = 1$. For any $f \in \mathscr{F}$

$$\phi_k(f) = \sum_{j_1} r_{j_1}^s \sum_{j_2\ldots j_k} (r_{j_2}\ldots r_{j_k})^s f(x_{j_1\ldots j_k})$$

$$= \sum_{j_1} r_{j_1}^s \sum_{j_2\ldots j_k} (r_{j_2}\ldots r_{j_k})^s f(\psi_{j_1}(x_{j_2\ldots j_k}))$$

$$= \sum_j r_j^s \phi_{k-1}(f \circ \psi_j).$$

Letting $k \to \infty$ we get

$$\int f\,\mathrm{d}\mu = \sum_j r_j^s \int f \circ \psi_j\,\mathrm{d}\mu$$

if $f \in \mathscr{F}$, using (8.17). By the usual approximation process using the monotone convergence theorem this also holds for all non-negative functions f, so (8.15) follows.

Finally, if f is any continuous function vanishing on E, we have $\phi_k(f) = 0$ for all k by (8.16), so $\int f\,\mathrm{d}\mu = 0$ by (8.17). Thus μ is supported by E. $\square$

It follows that, writing

$$\mu_{j_1\ldots j_k}(F) = \mu((\psi_{j_1}\circ\ldots\circ\psi_{j_k})^{-1}(F)) = \mu(\psi_{j_k}^{-1}\circ\ldots\circ\psi_{j_1}^{-1}(F)),$$

the measure $\mu_{j_1\ldots j_k}$ has support contained in $E_{j_1\ldots j_k}$. Also, from (8.15),

$$\mu_{j_1\ldots j_k} = \sum_j r_j^s \mu_{j_1\ldots j_k j}. \tag{8.18}$$

We say that the *open set condition* holds for the contractions $\{\psi_j\}_1^m$ if there exists a bounded open set V such that

$$\psi(V) = \bigcup_{j=1}^{m} \psi_j(V) \subset V \tag{8.19}$$

with this union disjoint. Then transforming by $\psi_{j_1 \ldots j_k}$,

$$\bigcup_{j=1}^{m} V_{j_1 \ldots j_k j} \subset V_{j_1 \ldots j_k},$$

again with a disjoint union. Thus the sets $\{V_{j_1 \ldots j_k}\}$ (with k arbitrary) form a net in the sense that any pair of sets from the collection are either disjoint or else have one included in the other.

If (8.19) holds then $E \subset \bar{V}$ (the bar denoting closure), indeed

$$E = \bigcap_{k=0}^{\infty} \psi^k(\bar{V}), \tag{8.20}$$

since $\{\psi^k(\bar{V})\}_k$ is a decreasing sequence of compact sets convergent to E in the Hausdorff metric by Theorem 8.3, which is impossible if E has points outside $\bar{V}$. Taking images under $\psi_{j_1} \circ \ldots \circ \psi_{j_k}$ we have $E_{j_1 \ldots j_k} \subset \bar{V}_{j_1 \ldots j_k}$. (Of course, since the ψ_j are continuous $\psi_j(\bar{V}) \subset \overline{\psi_j(V)}$ etc.)

We now assume that the $\{\psi_j\}$ are similitudes. We show that if the open set condition holds, then the invariant set E is self-similar and the Hausdorff dimension and similarly dimension of E are equal. The *similarity dimension*, which has the advantage of being easily calculable, is the unique positive number s for which

$$\sum_{1}^{m} r_j^s = 1.$$

The open set condition ensures that the sets $\psi_j(E)$ cannot overlap too much.

Lemma 8.5

Let $\{V_i\}$ be a collection of disjoint open subsets of $\mathbb{R}^n$ such that each V_i contains a ball of radius $c_1 \rho$ and is contained in a ball of radius $c_2 \rho$. Then any ball B of radius ρ intersects, at most, $(1 + 2c_2)^n c_1^{-n}$ of the sets $\bar{V}_i$.

Proof. If $\bar{V}_i$ meets B, then $\bar{V}_i$ is contained in a ball concentric with B and of radius $(1 + 2c_2)\rho$. If q of the $\{\bar{V}_i\}$ meet B, then summing the volumes of the corresponding interior balls, $q(c_1 \rho)^n \leq (1 + 2c_2)^n \rho^n$, giving the stated bound for q. $\square$

Theorem 8.6

Suppose the open set condition holds for the similitudes ψ_j with ratios $r_j (1 \leq j \leq m)$. Then the associated compact invariant set E is an s-set, *where s is determined by*

$$\sum_1^m r_j^s = 1;\tag{8.21}$$

in particular $0 < \mathcal{H}^s(E) < \infty$.

Proof. Iterating (8.13) $E = \bigcup_{j_1 \ldots j_k} E_{j_1 \ldots j_k}$. By (8.14)

$$\sum_{j_1 \ldots j_k} |E_{j_1 \ldots j_k}|^s = \sum_{j_1 \ldots j_k} |E|^s (r_{j_1} \ldots r_{j_k})^s = |E|^s.$$

As $|E_{j_1 \ldots j_k}| \leq (\max_j r_j)^k |E| \to 0$ as $k \to \infty$ we conclude that $\mathcal{H}^s(E) \leq |E|^s < \infty$.

The lower bound is obtained using the two lemmas. Suppose that the open set V for which (8.19) holds contains a ball of radius c_1 and is contained in a ball of radius c_2. Take any $\rho > 0$. For each infinite sequence $\{j_1, j_2, \ldots\}$ with $1 \leq j_i < m$, curtail the sequence at the least value of $k \geq 1$ for which

$$(\min_j r_j)\rho \leq r_{j_1} \ldots r_{j_k} \leq \rho \tag{8.22}$$

and let $\mathscr{S}$ denote the set of finite sequences obtained in this way. It follows from the net property of the open sets that $\{V_{j_1 \ldots j_k} : j_1 \ldots j_k \in \mathscr{S}\}$ is a disjoint collection. Each such $V_{j_1 \ldots j_k}$ contains a ball of radius $c_1 r_{j_1} \ldots r_{j_k}$ and hence one of radius $c_1(\min_j r_j)\rho$, by (8.22), and similarly is contained in a ball of radius $c_2 r_{j_1} \ldots r_{j_k}$ and so in a ball of radius $c_2\rho$. By Lemma 8.5 any ball B of radius ρ intersects, at most, $q = (1 + 2c_2)^n c_1^{-n}(\min_j r_j)^{-n}$ sets of the collection $\{V_{j_1 \ldots j_k} : j_1 \ldots j_k \in \mathscr{S}\}$. Also $\mu_{j_1 \ldots j_k}(\mathbb{R}^n) = 1$ and $\text{support}(\mu_{j_1 \ldots j_k}) \subset E_{j_1 \ldots j_k} \subset \overline{V}_{j_1 \ldots j_k}$ for any $\{j_1, \ldots, j_k\}$. Iterating (8.18) as appropriate we see that

$$\mu = \sum_{j_1 \ldots j_k \in \mathscr{S}} (r_{j_1} \ldots r_{j_k})^s \mu_{j_1 \ldots j_k},$$

so that $\mu(B) \leq \sum (r_{j_1} \ldots r_{j_k})^s \mu_{j_1 \ldots j_k}(\mathbb{R}^n)$, where the sum is over those sequences $\{j_1, \ldots, j_k\}$ in $\mathscr{S}$ for which $\overline{V}_{j_1 \ldots j_k}$ intersects B. Thus, using (8.22), $\mu(B) \leq q\rho^s = q2^{-s}|B|^s$ for any ball with $|B| < |V|$. But, given any cover $\{U_i\}$ of E, we may cover E by balls $\{B_i\}$ with $|B_i| \leq 2|U_i|$, so

$$1 = \mu(E) \leq \Sigma \mu(B_i) \leq q2^{-s}\Sigma |B_i|^s \leq q\Sigma |U_i|^s.$$

We may choose $\{U_i\}$ to make $\Sigma |U_i|^s$ arbitrarily close to $\mathcal{H}^s(E)$, so $\mathcal{H}^s(E) \geq q^{-1} > 0$, as required. $\quad\square$

Corollary 8.7
If the open set condition holds, then $\mathcal{H}^s(\psi_i(E) \cap \psi_j(E)) = 0$ $(i \neq j)$, *so, in particular, E is self-similar.*

Proof. Using the fact that the ψ_j are similitudes

$$\sum_{j=1}^{m} \mathscr{H}^s(\psi_j(E)) = \sum_{j=1}^{m} r_j^s \mathscr{H}^s(E) = \mathscr{H}^s(E).$$

As $0 < \mathscr{H}^s(E) < \infty$ this can only happen if $\mathscr{H}^s(\psi_i(E) \cap \psi_j(E)) = 0$ ($i \neq j$), using (8.13) and the additive properties of $\mathscr{H}^s$.

This technique may be used to estimate Hausdorff dimensions if the ψ_j are contractions rather than similarities. We give one useful case here (see also Exercise 8.5).

Theorem 8.8

Let $\{\psi_j\}_1^m$ be contractions on $\mathbb{R}$ for which the open set condition (8.19) holds with V an open interval. Suppose that for each j,

$$q_j|x - y| \le |\psi_j(x) - \psi_j(y)| \le r_j|x - y| \qquad (8.23)$$

for all $x, y \in \bar{V}$. Then $s \le \dim E \le t$, where s and t are defined by $\sum_1^m q_j^s = 1 = \sum_1^m r_j^t$.

Proof. The method of Theorem 8.6 gives the upper bound. To obtain the lower bound we proceed as there, but instead of using Lemma 8.5 we use the fact that if $\{V_i\}$ are a collection of disjoint *intervals*, each of length at least $2c_1\rho$, then any interval B of length 2ρ intersects, at most, $c_1^{-1} + 2$ of the $\bar{V}_i$. $\quad \square$

When the ψ_j are merely contractions, it is usually possible to obtain better estimates for $\dim E$ by working with the mappings $\{\psi_{j_1 \ldots j_k}\}$ for k larger than 1 and replacing (8.23) by tighter estimates.

The methods of Section 1.5 may also be used to estimate the dimensions of such sets.

Mandelbrot (1982) has introduced the notion of the generator of a self-similar set which conveniently specifies a set of similitudes. A generator consists of a number of straight-line segments and two points specially identified. We associate with each line segment the similitude which maps the two special points onto the endpoints of the segment. Using the generator a sequence of sets approximating to the self-similar set may be built up by iterating the process of replacing each line segment by a similar copy of the generator. The sequences of sets obtained in this way are the sets $\psi^k(G)$, where G is the generator. Note that in $\mathbb{R}^2$ the similitudes are defined only to within a reflection, but the orientations may be prescribed by showing the first iteration. The dimension of the set may easily be calculated from the generator using Theorem 8.6. The idea is best illustrated by examples. Figures 8.1(*b*) and 8.2(*b*) show the generators and the first few stages of approximation for the curves in Figures 8.1(*a*) and 8.2(*a*).

As a variant, one may introduce an element of randomness into these constructions. It is possible to develop a theory if the ψ_j are selected at random according to a certain probability distribution at each step of the iterated process. In this case the sets obtained will not themselves be self-similar, but will be *statistically self-similar* in the sense that the sets will have the same probability distribution as their component subsets. Under reasonable conditions such sets achieve, with probability 1, a particular dimension which may be expressed in terms of the expected values of the ratios of the ψ_j.

8.4 Osculatory and Apollonian packings

If V is a bounded open subset of $\mathbb{R}^n$, an *osculatory packing* of V is a sequence of closed balls $\{B_i\}_1^\infty$ such that, for each j, B_j is the largest ball (or one of the largest) contained in $V \backslash \bigcup_1^{k-1} B_i$. The most interesting case, called the *Apollonian packing*, is obtained if V is the interior of a curvilinear triangle in $\mathbb{R}^2$ (see Figure 8.3 for the first few stages of this packing).

Of particular interest is the Hausdorff dimension of the closed set $E = V \backslash \bigcup_1^\infty B_i$, known as the *residual set* of the packing. It is obvious that $1 \leq \dim E \leq 2$ for the Apollonian packing, but it is surprisingly difficult to

Fig. 8.3

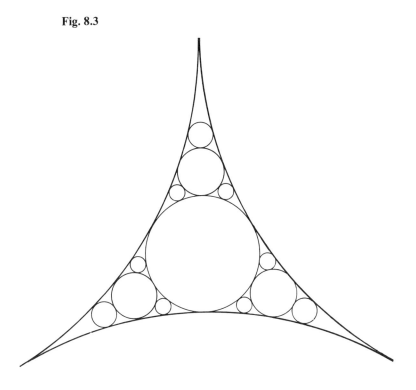

show that these inequalities are strict, let alone to obtain good estimates for dim E. Hirst (1967) was the first to obtain non-trivial estimates, and we outline his arguments here, with some refinements due to Boyd (1973b). We omit the details of some of the tedious but straightforward algebraic computations.

The following formulae for the inradius and circumradius of a curvilinear triangle may be proved using elementary trigonometry. The expression for the inradius, known as Soddy's formula, is proved in Coxeter (1961, p. 12) and was expressed in rhyme by the physicist Soddy (1936).

Lemma 8.9
Let T be a curvilinear triangle formed by externally touching circles of radii a, b and c. The inradius r and circumradius ρ of T are given by

$$r^{-1} = a^{-1} + b^{-1} + c^{-1} + 2(b^{-1}c^{-1} + a^{-1}c^{-1} + a^{-1}b^{-1})^{\frac{1}{2}},$$
$$\rho^2 = (abc)/(a+b+c) = (b^{-1}c^{-1} + a^{-1}c^{-1} + a^{-1}b^{-1})^{-1}.$$

Let T be any curvilinear triangle, with sides of radii a, b and c and with B in its inscribed disc. Let T_1, T_2 and T_3 be the three curvilinear triangles that remain on removing B from T. Suppose that T, T_j have circumradii ρ, ρ_j.

Lemma 8.10
If $s \geq \log 3/\log(1 + 2 \cdot 3^{-\frac{1}{2}}) = 1.4311\ldots$, then $\rho^s \geq \rho_1^s + \rho_2^s + \rho_3^s$.

Proof. Applying Lemma 8.9 to T and T_1 we get, after some calculation,

$$\rho/\rho_1 = 1 + \rho(b^{-1} + c^{-1})$$

(where the sides common to T and T_1 have radii b and c) with similar formulae for T_2 and T_3. Hence

$$\rho_1^s + \rho_2^s + \rho_3^s = \rho^s[(1 + (b^{-1} + c^{-1})\rho)^{-s} + (1 + (a^{-1} + c^{-1})\rho)^{-s} + (1 + (a^{-1} + b^{-1})\rho)^{-s}]$$

with $\rho = (b^{-1}c^{-1} + a^{-1}c^{-1} + b^{-1}c^{-1})^{-\frac{1}{2}}$. The expression in square brackets is homogeneous in a, b and c, so in order to show that it is bounded above by 1 for any a, b and c we may assume that $\rho = 1$. This is a problem in optimization under constraint and it is not difficult to obtain the desired bound provided that $s \geq \log 3/\log(1 + 2 \cdot 3^{-\frac{1}{2}})$, see Hirst (1967). $\quad\square$

Now let T be an equilateral curvilinear triangle with sides of unit radius. Removing the closed inscribed disc B from T we are left with three curvilinear triangles T_1, T_2 and T_3. Removing the inscribed discs B_1, B_2 and B_3 of these three triangles leaves nine curvilinear triangles $T_{j_1 j_2}(1 \leq j_1, j_2 \leq 3)$ with inscribed discs $B_{j_1 j_2}$. We proceed in this way, at each stage removing the inscribed disc $B_{j_1 \ldots j_k}$ from $T_{j_1 \ldots j_k}$ to give the curvilinear triangles $T_{j_1\ldots j_k1}, T_{j_1\ldots j_k2}, T_{j_1\ldots j_k3}$. (We allocate the indices 1, 2, 3 consistently to indicate the position of the triangles relative to $B_{j_1\ldots j_k}$.) Observe that the interiors of the curvilinear triangles thus obtained form a net of sets. The collection of discs $\{B_{j_1\ldots j_k}\}$ for all finite sequences $\mathbf{j} = \{j_1,\ldots,j_k\}$ with $1 \le j_i \le 3$ gives the Apollonian packing of T.

Theorem 8.11

If E is the residual set for the Apollonian packing then $\dim E \le \log 3/ \log(1 + 2\cdot3^{-\frac{1}{2}}) = 1.4311\ldots$.

Proof. Given $\delta > 0$ we may choose k large enough to ensure that $|T_{\mathbf{j}}| \le \delta$ for all sequences $\mathbf{j}$ of length k. Since $E \subset \bigcup_{\mathbf{j}} T_{\mathbf{j}}$, where the union is over all such k-tuples,

$$\mathcal{H}^s_\delta(E) \le \sum_{j_1\ldots j_k} |T_{j_1\ldots j_k}|^s \le 2^s \sum_{j_1\ldots j_k} \rho^s_{j_1\ldots j_k},$$

where $\rho_{\mathbf{j}}$ is the circumradius of $T_{\mathbf{j}}$. If $s > \log 3/\log(1 + 2\cdot3^{-\frac{1}{2}})$, then by repeated applications of Lemma 8.10

$$\sum_{j_1\ldots j_k} \rho^s_{j_1\ldots j_k} \le \sum_{j_1\ldots j_{k-1}} \rho^s_{j_1\ldots j_{k-1}} \le \cdots \le \sum_{j_1} \rho^s_{j_1} \le \rho^s,$$

where $\rho = 3^{-\frac{1}{2}}$ is the circumradius of T. Thus $\mathcal{H}^s_\delta(E) \le (2\cdot3^{-\frac{1}{2}})^s$ for all $\delta > 0$, giving $\mathcal{H}^s(E) \le (2\cdot3^{-\frac{1}{2}})^s$. □

To obtain a lower bound for the dimension of the residual set we estimate the inradii $r_{\mathbf{j}}$ of the curvilinear triangles $T_{\mathbf{j}}$. Denote the curvatures (that is, the reciprocals of the radii) of the sides of $T_{\mathbf{j}}$ by $\alpha_{\mathbf{j}}, \beta_{\mathbf{j}}, \gamma_{\mathbf{j}}$, again with a consistent convention for orientation, and let

$$\sigma_{\mathbf{j}} = (\beta_{\mathbf{j}}\gamma_{\mathbf{j}} + \alpha_{\mathbf{j}}\gamma_{\mathbf{j}} + \alpha_{\mathbf{j}}\beta_{\mathbf{j}})^{\frac{1}{2}}. \tag{8.24}$$

We may, in principle at least, calculate these radii and curvatures by repeated applications of Soddy's formula. Let M_1, M_2 and M_3 be the matrices

$$M_1 = \begin{vmatrix} 1 & 0 & 0 & 0 \\ 1 & 1 & 0 & 1 \\ 1 & 0 & 1 & 1 \\ 2 & 0 & 0 & 1 \end{vmatrix}, \quad M_2 = \begin{vmatrix} 1 & 1 & 0 & 1 \\ 0 & 1 & 0 & 0 \\ 0 & 1 & 1 & 1 \\ 0 & 2 & 0 & 1 \end{vmatrix}, \quad M_3 = \begin{vmatrix} 1 & 0 & 1 & 1 \\ 0 & 1 & 1 & 1 \\ 0 & 0 & 1 & 0 \\ 0 & 0 & 2 & 1 \end{vmatrix}.$$

By Lemma 8.9 it follows that the row vectors $(\alpha_{\mathbf{j}}, \beta_{\mathbf{j}}, \gamma_{\mathbf{j}}, \sigma_{\mathbf{j}})$ are related by

$$(\alpha_{\mathbf{j}p}, \beta_{\mathbf{j}p}, \gamma_{\mathbf{j}p}, \sigma_{\mathbf{j}p}) = (\alpha_{\mathbf{j}}, \beta_{\mathbf{j}}, \gamma_{\mathbf{j}}, \sigma_{\mathbf{j}})M_p \quad (p = 1, 2, 3),$$

where $\mathbf{j}p$ denotes the sequence $\{j_1,\ldots,j_k,p\}$. (It is easy to check that (8.24) then holds with $\mathbf{j}$ replaced by $\mathbf{j}p$.) Hence, iterating,

$$(\alpha_{\mathbf{j}}, \beta_{\mathbf{j}}, \gamma_{\mathbf{j}}, \sigma_{\mathbf{j}}) = (1, 1, 1, \sqrt{3})M_{j_1}M_{j_2}\ldots M_{j_k}, \tag{8.25}$$

enabling us to find the radius of B_j as

$$r_j^{-1} = \alpha_j + \beta_j + \gamma_j + 2\sigma_j = (1, 1, 1, \sqrt{3}) M_{j_1} M_{j_2} \ldots M_{j_k} (1, 1, 1, 2)', \qquad (8.26)$$

where the prime denotes the transpose of a vector.

Thus the rates at which the radii of the packing circles tend to zero depend directly on the behaviour of the matrix products $M_{j_1} M_{j_2} \ldots M_{j_k}$ as $k \to \infty$. The matrices in this product may be regarded as selected at random from M_1, M_2 and M_3 with equal probability and the information we require is then contained in the probability distribution of the entries in the matrix product for large k. A theory of products of random matrices is being developed, see Kingman (1973, 1976), and we indicate a method of estimating dim E from below using these ideas rather than the more complex but computationally more efficient methods of Hirst (1967) and Boyd (1973a). It may be shown using a modification of the proof of Kingman (1976), Theorem 2.2, that there is a number λ such that the kth root of every entry of the matrix $M_{j_1} M_{j_2} \ldots M_{j_k}$ converges to λ as $k \to \infty$ for almost every sequence $\{j_1, j_2, \ldots\}$ $(1 \leq j_i \leq 3)$ with the natural product measure on sets of infinite sequences. The important thing is that with the matrices M_1, M_2 and M_3 as given, $1 < \lambda < 3$. Thus for any $\varepsilon > 0$ we may find m large enough to ensure that for a proportion $1 - \varepsilon$ of all sequences $\{j_1, \ldots, j_m\}$ of length m all entries of $M_{j_1} \ldots M_{j_m}$ are less than $(\lambda + \varepsilon)^m$. Of course, this fact may be verified by direct calculation. For a given ε, let $\mathcal{U}$ denote the set of m-term sequences with this property.

Lemma 8.12
Suppose $t \leq (\log 3 + m^{-1} \log(1 - \varepsilon))/\log(\lambda + \varepsilon)$. Let $\mathcal{S}$ be a finite set of finite sequences such that the triangles $\{T_j : j \in \mathcal{S}\}$ cover the residual set E. Then

$$\sum_{j \in \mathcal{S}} r_j^t \geq c, \qquad (8.27)$$

where c is a positive constant.

Proof. It is enough to prove the lemma on the assumption that each sequence in $\mathcal{S}$ has a multiple of m terms. For, replacing each T_j by $T_{j'}$, where j' consists of the first mq terms of j where q is the quotient on dividing the number of terms in j by m, increases each term of the sum in (8.27) by a factor of, at most, 5^{mt}, using (8.26). By the net property of the triangles we may further assume that $\{T_j : j \in \mathcal{S}\}$ is a disjoint collection except for common vertices. Let $j_1, j_2, \ldots, j_q$ be one of the longest sequences in $\mathcal{S}$, where each j_i is an m-tuple. Then, since the triangles cover E, the sequence

$\mathbf{j}_1, \mathbf{j}_2, \ldots, \mathbf{j}_{q-1}, \mathbf{j}$ must be in $\mathscr{S}$ for every m-tuple $\mathbf{j}$. But if $\mathbf{j} \in \mathscr{U}$, then, by (8.26),

$$r^{-1}_{\mathbf{j}_1 \ldots \mathbf{j}_{q-1} \mathbf{j}} = (1,1,1,\sqrt{3}) M_{\mathbf{j}_1} \ldots M_{\mathbf{j}_{q-1}} M_{\mathbf{j}} (1,1,1,2)'$$
$$\leq (1,1,1,\sqrt{3}) M_{\mathbf{j}_1} \ldots M_{\mathbf{j}_{q-1}} (1,1,1,2)' (\lambda + \varepsilon)^m$$
$$\leq (\lambda + \varepsilon)^m r^{-1}_{\mathbf{j}_1 \ldots \mathbf{j}_{q-1}}.$$

Thus with t as stated,

$$\sum_{\mathbf{j}} r^t_{\mathbf{j}_1 \ldots \mathbf{j}_{q-1} \mathbf{j}} \geq \sum_{\mathbf{j} \in \mathscr{U}} r^t_{\mathbf{j}_1 \ldots \mathbf{j}_{q-1} \mathbf{j}} \geq (1-\varepsilon) 3^m (\lambda + \varepsilon)^{-mt} r^t_{\mathbf{j}_1 \ldots \mathbf{j}_{q-1}}$$

$$\geq r^t_{\mathbf{j}_1 \ldots \mathbf{j}_{q-1}},$$

so replacing the triangles $T_{\mathbf{j}_1 \ldots \mathbf{j}_{q-1} \mathbf{j}}$ by the single triangle $T_{\mathbf{j}_1 \ldots \mathbf{j}_{q-1}}$ does not increase the sum in (8.27). We may repeat this process until we reach the single triangle T when (8.27) is obvious. Hence (8.27) holds for any cover of E by a subcollection of the curvilinear triangles. $\square$

Lemma 8.13
Let $\{V_i\}$ be a finite set of open discs with $E \subset \bigcup_i V_i$. Then there is a finite set of sequences $\mathscr{S}$ such that $E \subset \bigcup_{\mathbf{j} \in \mathscr{S}} T_{\mathbf{j}}$ and

$$\sum_i |V_i|^t \geq 4^{-t} \sum_{\mathbf{j} \in \mathscr{S}} r^t_{\mathbf{j}}. \tag{8.28}$$

Proof. Clearly, we may assume that each V_i overlaps T and also that none of the discs contains any other. We consider each disc V_i in turn. Let k be the greatest integer such that V_i meets just one triangle $T_{\mathbf{j}}$ with the sequence $\mathbf{j}$ containing k terms. We examine two cases separately.
(a) If V_i meets $T_{\mathbf{j}1}, T_{\mathbf{j}2}$ and $T_{\mathbf{j}3}$, then V_i must meet at least three of the four circles forming the boundaries of these triangles, so the radius of V_i is at least the inradius of the curvilinear triangle formed by these three circles. Thus

$$\tfrac{1}{2}|V_i| \geq \min\{r_{\mathbf{j}1}, r_{\mathbf{j}2}, r_{\mathbf{j}3}, r_{\mathbf{j}}\} \geq \tfrac{1}{8} r_{\mathbf{j}}, \tag{8.29}$$

noting that the radii of the sides of $T_{\mathbf{j}}$ are greater than $r_{\mathbf{j}}$ and using Lemma 8.9, and we put $\mathbf{j}$ in $\mathscr{S}$.
(b) Suppose V_i meets just two of the triangles $T_{\mathbf{j}1}, T_{\mathbf{j}2}$ and $T_{\mathbf{j}3}$, without loss of generality $T_{\mathbf{j}1}$ and $T_{\mathbf{j}2}$. Let w be the point common to $T_{\mathbf{j}1}$ and $T_{\mathbf{j}2}$. If $w \notin V_i$, then, just as in (a),

$$\tfrac{1}{2}|V_i| \geq r_{\mathbf{j}3} \geq \tfrac{1}{8} r_{\mathbf{j}} \tag{8.30}$$

and we put $\mathbf{j}$ in $\mathscr{S}$. Otherwise, $w \in V_i$. Write $B^0 = B_{\mathbf{j}1}$, $B^1 = B_{\mathbf{j}12}$, $B^2 = B_{\mathbf{j}122}, B^3 = B_{\mathbf{j}1222}, \ldots$. Let m be the least integer such that B^m meets V_i. Then $B^{m+1} \subset V_i$ so, using Lemma 8.9 as before,

$$|V_i| \geq |B^{m+1}| \geq \tfrac{1}{8}|B^m|. \tag{8.31}$$

Let $\mathbf{j}' \in \mathscr{S}$, where $\mathbf{j}' = \mathbf{j}1222\ldots2$ is the sequence corresponding to B^m. By a symmetrical argument applied to the balls of the form $B_{\mathbf{j}211\ldots1}$ we find a sequence $\mathbf{j}''$ corresponding to a triangle contained in $T_{\mathbf{j}2}$ which we also put in $\mathscr{S}$. Then $E \cap V_i \subset T_{\mathbf{j}'} \cup T_{\mathbf{j}''}$.

We observe that with this construction $E \subset \bigcup_{\mathbf{j} \in \mathscr{S}} T_{\mathbf{j}}$, with each $E \cap V_i$ covered by one or two of the triangles $\{T_{\mathbf{j}} : \mathbf{j} \in \mathscr{S}\}$. Combining (8.29)–(8.31) we get (8.28). $\square$

Theorem 8.14
For the Apollonian packing of the curvilinear equilateral triangle, $\dim E > 1$, *where E is the residual set of the packing.*

Proof. Choose ε and m above so that $t = (\log 3 + m^{-1} \log(1 - \varepsilon))/\log(\lambda + \varepsilon) > 1$. Then, combining Lemmas 8.12 and 8.13, we see that $\sum_i |V_i|^t \geq 4^{-t}c$ for any finite cover of E by open discs $\{V_i\}$. If $\{U_i\}$ is any cover of E by arbitrary sets we deduce that $\sum_i |U_i|^t \geq 2^{-t}4^{-t}c$ by enclosing each U_i in an open disc V_i and using the compactness of E. Thus $\mathscr{H}^t(E) > 0$, as required. $\square$

Of course, it follows that $\dim E \geq \log 3/\log \lambda$, where λ is the limiting number associated with the matrix products.

Hirst (1967) shows that $\dim E > 1.001$ by estimating the effects of the products of just four matrices $M_{j_1} \ldots M_{j_4}$ in a rather more careful way than above. Boyd (1973a) has developed numerical methods to find $\dim E$ to any desired accuracy, and it is now known that $1.300 < \dim E < 1.315$. Larman (1967b) adopts a very different method to prove that the dimension of the residual set of any packing of a region by discs is at least 1.03.

Note that the Apollonian packings of any (concave-sided) curvilinear triangles are equivalent under a Möbius transformation of the form $z \to (az + b)/(cz + d)$ in complex notation. This is because such mappings, which may be chosen to map any curvilinear triangle onto any other, have the property of transforming circles to circles or, exceptionally, to straight lines. In particular, it follows from Lemma 1.8 that the residual sets of the Apollonian packings of all curvilinear triangles are of equal dimension. Further, we may find Möbius transformations ψ_1, ψ_2, ψ_3 which map T onto T_1, T_2 and T_3 respectively, using our earlier notation. Then the residual set E is a compact set such that $E = \bigcup_{j=1}^3 \psi_j(E)$. Thus although the ψ_j are just non-expansive rather than contractive, ideas from Section 8.3 may nevertheless be adapted to the study of Apollonian packings. A generalization of this problem is to find the Hausdorff dimension of the set of limit points of all mappings of the form $\psi_{j_1} \circ \ldots \circ \psi_{j_k}$, where the ψ_{j_i} are chosen from a given set of Möbius transformations, see Patterson (1976), Sullivan (1979) and Mandelbrot (1983).

An important number associated with any osculatory packing $\{B_i\}$ is the *exponent* of the packing e defined as the infimum value of t for which $\sum_{i=1}^{\infty} |B_i|^t < \infty$. It turns out that for the Apollonian packing the exponent e equals the dimension of the residual set E. It is relatively easy to show that $\dim E \le e$ (see Exercise 8.6; Larman (1966) gives a more general result.) The opposite inequality proved by Boyd (1973b) is much harder. Careful consideration of the matrix products is required to show that if $t < e$ there exists a constant c such that if the curvilinear triangles $\{T_\mathbf{j} : \mathbf{j} \in \mathscr{S}\}$ cover E, then $\sum_{\mathbf{j} \in \mathscr{S}} r_\mathbf{j}^t \ge c$. The result then follows from Lemma 8.13.

Eggleston (1953a) studies the residual set resulting from packing an equilateral triangle with smaller equilateral triangles of opposite orientation. The configuration indicated in Figure 8.4 leaves a residual set of dimension $\log 3/\log 2$, which is the smallest value obtainable for any such packing (see Exercise 8.7).

Fig. 8.4

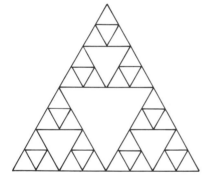

8.5 An example from number theory

Hausdorff measures have found many applications in the theory of numbers, particularly in Diophantine approximation. Typically, 'the set of numbers approximable to a specified accuracy by rationals cannot have too large Hausdorff dimension'. Here we consider particular results of this type which provide an instructive example in connection with the projection theorems of Chapter 6.

As is usual in number theory we let $\|x\|$ denote the distance of the real number x from the nearest integer. We require both the one- and two-dimensional versions of the following theorem.

Theorem 8.15

Take $\alpha > 0$ and a sequence of positive integers $n_1, n_2, \ldots$ such that

$$n_{j+1} \geq n_j^j. \tag{8.32}$$

(a) *Let F be the set of real x such that*

$$\|n_j x\| \leq n_j^{-\alpha} \quad (j = 1, 2, \ldots). \tag{8.33}$$

Then $\dim F = 1/(1 + \alpha)$.

(b) *Let E be the set of $(x, y) \in \mathbb{R}^2$ such that*

$$\|n_j x\|, \|n_j y\| \leq n_j^{-\alpha} \quad (j = 1, 2, \ldots).$$

Then $\dim E = 2/(1 + \alpha)$.

Proof. (a) Writing

$$F_j = \bigcup_{i = -\infty}^{\infty} [in_j^{-1} - n_j^{-1-\alpha}, in_j^{-1} + n_j^{-1-\alpha}], \tag{8.34}$$

then

$$F = \bigcap_{j=1}^{\infty} F_j. \tag{8.35}$$

First note that the intersection of F with any interval I may be covered by, at most, $|I|n_j + 2$ intervals of length $2n_j^{-1-\alpha}$, for every j. Since these lengths tend to 0 as j tends to infinity, the set $F \cap I$ has finite $\mathscr{H}^{1/(1+\alpha)}$-measure, so $\dim F \leq 1/(1 + \alpha)$.

To find a lower bound for the dimension, we take $s < 1/(1 + \alpha)$ and show, by modifying F slightly to fit Theorem 1.15, that $\mathscr{H}^s(F) > 0$. We may assume that $n_{j+1}n_j^{-1-\alpha} \geq 2$ for all j, that $n_j^\alpha \geq 4$ and that $n_1 \geq 4$; if this is not so we may ignore small values of j and renumber, using (8.32). Let

$$m_j = [2n_{j+1}n_j^{-1-\alpha} - 2] \geq n_{j+1}n_j^{-1-\alpha} \geq 2, \tag{8.36}$$

where [ ] denotes 'greatest integer not more than'. Observe that we may find $c > 0$ such that

$$m_1 \ldots m_{j-1} n_j^{-s(1+\alpha)} \geq c \quad (j \geq 2). \tag{8.37}$$

For, on replacing j by $j + 1$, the left-hand side of this inequality increases by a factor

$$m_j n_{j+1}^{-s(1+\alpha)} n_j^{s(1+\alpha)} \geq n_{j+1}^{1-s(1+\alpha)} n_j^{(s-1)(1+\alpha)}$$
$$\geq n_j^{j[1-s(1+\alpha)]+(s-1)(1+\alpha)}$$

by (8.36) and (8.32), and this is greater than 1 if j is large enough, since $1 - s(1 + \alpha) > 0$.

Each interval of F_j contains either m_j, $m_j + 1$ or $m_j + 2$ complete intervals of F_{j+1} depending on the overlaps at either end. To allow for these end effects we reduce F to F' as follows. Let F_0' be the unit interval. If $j \geq 0$ and

F'_j has been specified, define F'_{j+1} by choosing exactly m_j consecutive complete intervals of F_{j+1} from each interval of F'_j. Let $F' = \bigcap_{j=0}^{\infty} F'_j$ so that $F' \subset F \cap [0, 1]$.

We now construct a subset E of $[0, 1]$ by the Cantor-like process described prior to Theorem 1.15. Let $E_0 = [0, 1]$ and let $E_1, E_2, \ldots$ be formed in the manner described there with each interval of E_j containing m_j intervals of E_{j+1}, so that (1.23) becomes

$$m_j |J|^s = |I|^s \tag{8.38}$$

for a typical interval J of E_{j+1} and an interval I of E_j. By Theorem 1.15 the dimension of the set $E = \bigcap_{j=0}^{\infty} E_j$ is s.

There is a natural bijection ψ from F' to E: if $x \in F'$ lies in the k_jth interval of F'_j (counting from the left), then $\psi(x)$ is the unique point that is in the k_jth interval of E_j for all j. Suppose that x and y are distinct points of F'. Let j be the least integer such that x and y lie in different intervals of F'_{j+1}; say that y lies in an interval k to the right of the one containing x. Then

$$|y - x| \geq kn_{j+1}^{-1} - 2n_{j+1}^{-1-\alpha} \geq \tfrac{1}{2} kn_{j+1}^{-1}. \tag{8.39}$$

On the other hand, the points $\psi(x)$ and $\psi(y)$ of E lie in the same interval I of E_j but with $\psi(y)$ in the interval of E_{j+1} lying k to the right of that of $\psi(x)$. Thus

$$|\psi(y) - \psi(x)| \leq 2k|I|m_j^{-1} \leq 2k(m_1 \ldots m_{j-1})^{-1/s} m_j^{-1}$$
$$\leq 2kc_1 n_j^{-1-\alpha} m_j^{-1} \leq 2kc_1 n_{j+1}^{-1},$$

where $c_1 = c^{-1/s}$, iterating (8.38) and using (8.37) and (8.36). Combining with (8.39)

$$|\psi(y) - \psi(x)| \leq 4c_1 |y - x|.$$

We conclude from Lemma 1.8 that $s = \dim E \leq \dim F' \leq \dim F$. This holds for all $s < 1/(1 + \alpha)$.

(b) One way of proving (b) would be to modify the proofs of Theorem 1.15 and (a) to two dimensions. Alternatively, the set $E = F \times F$ has dimension at least $2/(1 + \alpha)$, using Corollary 5.10 on Cartesian product sets together with part (a). On the other hand, If S is any square of unit side, we may cover $E \cap S$ by $(n_j + 2)^2$ squares of diameter $2^{\frac{1}{2}} n_j^{-1-\alpha}$ for each j (taking squares from $F_j \times F_j$). Thus $\dim(E \cap S)$, and hence $\dim E$, is, at most, $2/(1 + \alpha)$. □

Observe that the x satisfying (8.33) are those numbers 'closely approximable' in the sense that $|x - p_j/n_j| \leq n_j^{-1-\alpha}$ for some rational number p_j/n_j for all j.

If inequality (8.33) is required to hold 'for infinitely many j' rather than 'for all j', the corresponding result is harder to prove. In this case the intersection (8.35) is replaced by an upper limit. We state without proof the 1-dimensional version of Jarnik's theorem. Proofs, all rather technical, are given by Jarnik (1931), Besicovitch (1934b), Eggleston (1952) and Kaufman (1970, 1981), see also Good (1941) for an interesting relationship with continued fractions. Part (b) gives some indication of how rapidly the solutions increase.

Theorem 8.16
Take $\beta > 1$.
(a) *The set of real numbers x for which the inequality*

$$\|nx\| \leq n^{-\beta}$$

holds for infinitely many n has Hausdorff dimension $2/(1 + \beta)$.
(b) *Let $m_1, m_2, \ldots$ be a rapidly increasing sequence of integers, $m_{j+1} \geq m_j^j$, say. Then the set of real x for which*

$$\|nx\| \leq nm_j^{-1-\beta} \quad \text{and} \quad n \leq m_j$$

holds for some n for infinitely many j has Hausdorff dimension $2/(1 + \beta)$.

We now consider the projections of the plane set E of Theorem 8.15(b) onto lines. It turns out that the set of exceptional directions for the projections is as large as can possibly occur. (Compare Theorem 6.8.)

Theorem 8.17
Let $n_1, n_2 \ldots$ be a rapidly increasing sequence of positive integers, $n_{j+1} \geq n_j^j$, say. For $s \leq 1$ let E be the set of $(x, y) \in \mathbb{R}^2$ with $-1 \leq x, y \leq 1$ such that

$$\|n_j x\|, \|n_j y\| \leq n_j^{1-2/s} \quad (j = 1, 2, \ldots).$$

Then $\dim E = s$, and the set of directions

$$\{\theta : \dim (\text{proj}_\theta E) < s\} \tag{8.40}$$

has Hausdorff dimension at least s.

Proof. It is immediate from Theorem 8.15(b) that $\dim E = s$.
   For $\delta > 0$ let

$$A_\delta = \{a : \|na\| \leq nn_j^{-2/s} \quad \text{for some } n \text{ with}$$
$$1 \leq n \leq n_j^{1-\delta} \text{ for infinitely many } j\}.$$

Taking $m_j = n_j^{1-\delta}$ and $\beta = 2/s(1 - \delta) - 1$ in Theorem 8.16(b) gives

$$\dim A_\delta = s(1 - \delta). \tag{8.41}$$

For $a \in \mathbb{R}$, let $\psi_a : \mathbb{R}^2 \to \mathbb{R}$ be the linear mapping defined by

$$\psi_a(x, y) = ax + y. \tag{8.42}$$

If $a \in A_\delta$, then for arbitrarily large j, there exist integers n and b such that

$$1 \leq n \leq n_j^{1-\delta} \tag{8.43}$$

and

$$|na - b| \leq nn_j^{-2/s}$$

so that

$$|a - b/n| \leq n_j^{-2/s}. \tag{8.44}$$

If u, v are integers with $|u|, |v| \leq n_j$, then, in coordinate notation, write

$$(u, v) = (qn, - qb) + (r, v + qb),$$

where $u = qn + r$ with q, r chosen so that $0 \leq r < n$. If $a' = b/n$,

$$\psi_{a'}(u, v) = \psi_{a'}(qn, - qb) + \psi_{a'}(r, v + qb)$$
$$= 0 + \psi_{a'}(r, v + qb).$$

Since $(r, v + qb)$ takes, at most, cnn_j distinct values as (u, v) varies, where $c = 3(1 + |b|)$, the same is true of $\psi_{a'}(u, v)$ and thus of $\psi_{a'}(un_j^{-1}, vn_j^{-1})$. But every point of E has both coordinates within $n_j^{-2/s}$ of (un_j^{-1}, vn_j^{-1}) for some pair of integers u and v, so it follows that $\psi_{a'}(E)$ may be covered by cnn_j intervals of length $2(|a'| + 1)n_j^{-2/s}$. But by (8.44) $|\psi_a(x, y) - \psi_{a'}(x, y)| \leq n_j^{-2/s}$ if $|x|, |y| \leq 1$ so $\psi_a(E)$ may be covered by cnn_j intervals of length $2(|a'| + 2)n_j^{-2/s}$. We may do this for arbitraily large j so $\mathscr{H}^t(\psi_a(E)) = 0$, provided that $nn_j n_j^{-2t/s} \rightarrow 0$ as $j \rightarrow \infty$. This is so if $t > (1 - \frac{1}{2}\delta)s$ by (8.43).

Thus we have shown that $\dim(\psi_a(E)) < s$ if $a \in A_\delta$ for any $\delta > 0$, that is, if $a \in A = \bigcup_{\delta > 0} A_\delta$. By (8.41) $\dim A = s$. From the definition of ψ_a, if $a = \cot \theta$ the sets $\psi_a(E)$ and $\text{proj}_\theta(E)$ are geometrically similar and so have the same Hausdorff dimension. We conclude that $\dim(\text{proj}_\theta(E)) < s$ for a set of θ of dimension at least s. ☐

The above example and its extensions to higher dimensions are due to Kaufman & Mattila (1975) and supersede those of Marstrand (1954a) and Kaufman (1969). By results of Mattila (1975) on the maximum dimension of the exceptional set of directions, the set (8.40) has dimension exactly s. Number-theoretic examples such as these generally seem to exhibit the worst possible behaviour from the projection point of view.

Further applications of Hausdorff measure to number theory may be found in Eggleston (1952), Baker & Schmidt (1970), Rogers (1970) and Baker (1978). For wider accounts of Diophantine approximation see Baker (1975) or Schmidt (1980).

8.6 Some applications to convexity

This section describes some of the ways in which sets of Hausdorff measure arise in the geometry of convex sets. The first two results show that the irregularities of a convex surface cannot be too great. The third result is an application of one of the projection theorems.

If K is a (compact) convex body in $\mathbb{R}^3$ we say x is a *singular point* of ∂K, the surface of K, if K supports more than one tangent plane at x. The following theorem, which generalizes to higher dimensions, is due to Anderson & Klee (1952); Besicovitch (1963*a*) gives an alternative proof.

Theorem 8.18
Let K be a convex body in $\mathbb{R}^3$. Then the set of singular points of ∂K is contained in a countable union of rectifiable curves and so is of σ-finite $\mathcal{H}^1$-measure.

Proof. Let Π be a plane that misses K, and let $f : \Pi \to \partial K$ be the 'nearest point' mapping. Thus if $x \in \Pi$, then $f(x)$ is the point of K for which $|f(x) - x|$ is least; since K is convex this point is unique. If $x, y \in \Pi$, then the angles $< xf(x)f(y)$ and $< yf(y)f(x)$ of the (possibly skew) quadrilateral $x, f(x), f(y), y$ must be at least $\frac{1}{2}\pi$, since the segment $[f(x), f(y)]$ is contained in K and contains candidates for 'nearest points' to x and y. Thus f is distance-decreasing, i.e.

$$|f(x) - f(y)| \leq |x - y| \quad (x, y \in \Pi). \tag{8.45}$$

Let ∂K_Π denote the image of Π under f, that is, roughly speaking, those points of ∂K on the same side of K as Π. If z is a singular point of ∂K_Π, then it is easy to see that $f^{-1}(z)$ contains a straight-line segment. Indeed, if z is a singular point of conical type (with z the vertex of a circular cone containing K), then $f^{-1}(z)$ contains a disc. Let $\{L_i\}_1^\infty$ be the countable collection of segments lying in Π which have endpoints with rational coordinates with respect to some fixed pair of axes in Π. If z is a singular point of ∂K_Π, then $f^{-1}(z)$ contains a segment which must cut L_i for some i. Thus the singular points of ∂K_Π are contained in $\bigcup_i f(L_i)$. But using (8.45) we see that the subset $f(L_i)$ of ∂K is either a rectifiable curve, or a single point (see Section 3.2). Thus the singular points of ∂K_Π lie in a countable collection of rectifiable curves and so are of σ-finite $\mathcal{H}^1$-measure.

Finally, if $\Pi_1, \ldots, \Pi_4$ are planes containing the faces of a tetrahedron that encloses K, every singular point of ∂K is a singular point of ∂K_{Π_j} for some $j (1 \leq j \leq 4)$. Thus the singular points of ∂K form a σ-finite regular 1-set. $\square$

Incidentally, it is easy to see from the proof that the set of conical singularities of a convex body is, at most, countable. (If z is a conical singularity of ∂K, then $f^{-1}(z)$ contains a disc, but the plane Π can only contain countably many disjoint discs.)

Klee (1957) asked how large the set of *directions* of straight-line segments lying in a convex surface could be. McMinn (1960) solved this problem in three dimensions and here we give a simplified version of Besicovitch's (1963*b*) proof.

Theorem 8.19

Let K be a convex body in $\mathbb{R}^3$. Then the set of directions of the line segments contained in ∂K has σ-finite $\mathscr{H}^1$-measure. (The directions are regarded as a subset of the unit sphere.)

Proof. Let Π_1 and Π_2 be parallel planes which intersect ∂K in the convex curves Γ_1 and Γ_2. We first prove that the $\mathscr{H}^1$-measure of the (compact) set of directions of the segments in ∂K which cut both Γ_1 and Γ_2 is σ-finite. Define a function $f_1 : \Gamma_1 \to \Gamma_2$ as follows: Let $x \in \Gamma_1$. For every plane Π with Γ_1 and Γ_2 on the same side, which touches Γ_1 at x and also touches Γ_2, $\Pi \cap \Gamma_2$ is either a single point or a straight-line segment. The aggregate of these points or segments over all such Π is a single point or a closed subarc of Γ_2. Let $f_1(x)$ be either this point or the midpoint of the arc. Clearly, f_1 is a monotonic function and so is of bounded variation. Consequently, the function $g_1 : \Gamma_1 \to \mathbb{R}^3$ given by the vector difference $g_1(x) = f_1(x) - x$ is also of bounded variation. Projecting orthogonally onto Π_1, $\mathrm{proj}_{\Pi_1} \circ g_1 : \Gamma_1 \to \Pi_1$ is of bounded variation, so that $\mathrm{proj}_{\Pi_1}(g_1(\Gamma_1))$ is a set of finite $\mathscr{H}^1$-measure by Exercise 3.1. But $\mathrm{proj}_{\Pi_1}(g(\Gamma_1))$ is the set of direction cosines of the segments $\{[x, f_1(x)] : x \in \Gamma_1\}$, so the set of direction of such segments has finite $\mathscr{H}^1$-measure.

We define $f_2 : \Gamma_2 \to \Gamma_1$ in exactly the same way and deduce that the set of directions of the segments $\{[f_2(y), y] : y \in \Gamma_2\}$ also has finite $\mathscr{H}^1$-measure.

If $[x, y]$ is a segment in ∂K with $x \in \Gamma_1$ and $y \in \Gamma_2$, that is not of the form $[x, f_1(x)]$ or $[f_2(y), y]$, one or other of two possibilities must occur which we consider in turn. First x can lie on a straight line segment L_1 in Γ_1 and y on a parallel segment L_2 in Γ_2. The set of directions of the segments $\{[x, y] : x \in L_1, y \in L_2\}$ clearly has finite $\mathscr{H}^1$-measure. Since a convex curve can contain only countably many straight-line segments there are, at most, countably many such pairs of line segments. The second possibility is for x and y to be, singular points of Γ_1 and Γ_2 (that is, points supporting more than one tangent). This can occur for at most countably many $[x, y]$. Thus the set of directions of both types of 'exceptional' segments $[x, y]$ has σ-finite $\mathscr{H}^1$-measure. Taking all cases together we conclude that the set of directions of boundary segments cutting both Γ_1 and Γ_2 has σ-finite $\mathscr{H}^1$-measure.

Finally, taking all planes parallel to, and at rational distances from, the three coordinate planes, we obtain a countable collection of pairs of parallel planes such that every segment in ∂K intersects some pair of planes in the collection. Thus the set of directions of boundary segments lies in a countable collection of σ-finite sets and so is σ-finite. $\qquad\square$

Ewald, Larman & Rogers (1970) completed the investigation of the higher-dimensional analogues of this theorem by showing that if K is a convex body in $\mathbb{R}^n$ the measure of the set of orientations (in the $G_{n,k}$ sense) of the k-dimensional balls lying in the boundary of K has σ-finite $\mathcal{H}^{k(n-k-1)}$-measure.

A related problem, concerning the measure of the points on the surface of a convex body that lie in some boundary segment (or, more generally, k-dimensional ball) is discussed by Burton (1979). Larman (1971b) considers the measure of the union of the relative boundaries of the faces of a convex body.

A well-known result of Klee (1959) states that if K is a convex body in $\mathbb{R}^3$ with every plane section a polygon, then K must be a polytope, that is, the convex hull of a finite set of points. It was conjectured that if K is a convex body with *almost* every section in *almost* every direction a polygon, then K has, at most, countably many *extreme* points (boundary points not lying in the interior of any segment in K). This conjecture was disproved by Dalla & Larman (1980) who showed that the set of extreme points could have Hausdorff dimension as large as 1. We do not repeat their intricate construction here, but simply deduce from the projection theorems that this is the worst case that can occur.

Theorem 8.20

Let K be a convex body in $\mathbb{R}^3$ with almost every plane section a polygon. Then the set E of extreme points of K has Hausdorff dimension at most 1.

Proof. This is an easy consequence of Mattila's refinement of the projection theorems applied to the Borel set E (see the penultimate paragraph of Section 6.3). If $\dim E > 1$ and $0 < t < \dim E - 1$, then, for almost all unit vectors θ,

$$\mathcal{H}^t(E \cap \text{proj}_\theta^{-1} u) = \infty$$

for a set of u on L_θ of positive $\mathcal{L}^1$-measure, where proj_θ denotes projection onto L_θ, the line through the origin in direction θ. For such θ and u, the plane $\text{proj}_\theta^{-1} u$ intersects E in an infinite set. Any extreme point of K is certainly an extreme point of the plane convex set $K \cap \text{proj}_\theta^{-1} u$, so these plane sections cannot be polygonal. $\square$

8.7 Attractors in dynamical systems

At the time of writing, one of the growth areas in mathematics and mathematical physics is the study of 'chaos' and 'strange attractors' in dynamical systems. We attempt to compress some of the basic ideas of how such attractors arise into a few pages.

Let f be a mapping of a metric space (usually a subset of $\mathbb{R}^n, n \geq 1$) into itself. We are interested in the behaviour of the sequences of points or *orbits* $\{f^m(x)\}_{m=1}^\infty$ for various initial points x, particularly for large values of m.

(As usual, f^m denotes the mth iterate of f.) For example, the sequences may be periodic or may converge to a periodic orbit. Alternatively, $\{f^m(x)\}$ may appear to wander about the metric space almost at random. One is rapidly led to study *invariant* sets, that is, sets E for which $f(E) \subset E$. Then if $x \in E$ the iterates $f^m(x)$ remain trapped in E for all m. If f is continuous the closure of an invariant set is invariant, and a major problem is to find the closed invariant sets of a given function f. In cases of particular interest an invariant set may exhibit a fine structure. Perhaps the simplest example is the function $f : \mathbb{R} \to \mathbb{R}$ defined by $f(x) = \frac{3}{2}(1 - |2x - 1|)$ for which the Cantor set is invariant, see Exercise 8.8. Some remarkable sets known as Julia sets and defined by the requirement that $f(E) = E$, where f is the transformation of the complex plane $z \to z^2 - \mu$, are illustrated by Mandelbrot (1982, Section 19) for various values of the parameter μ.

However, the invariant sets of greatest practical importance have some sort of stability associated with them, otherwise they are unlikely to be observed in physical situations or in computer realizations. These sets are known as *attractors*. Various definitions of attractors have been given; essentially an attractor of a mapping f is a closed set E such that $f(E) \subset E$ and such that if $x \in V$, then the distance from $f^m(x)$ to E tends to 0 as $m \to \infty$, where V is a 'large' set containing E, for example an open neighbourhood of E. One usually also demands that E is minimal in some sense, perhaps by requiring that the orbit $\{f^m(x)\}$ be dense in E for some x. If E has a fine structure, or if there is sensitive dependence on the initial conditions (so that two nearby points may not remain close under iterates of f), E may be referred to as a strange attractor. Much work remains to be done to determine which functions give rise to strange attractors, why such attractors occur and what their structure is. The subject has become extremely complex with bifurcation theory, ergodic theory, differential topology and functional analysis all playing important rôles. For an excellent survey article on mappings in 1 and 2 dimensions see Whitley (1983).

Considerable progress has been made recently for mappings of a real interval, largely by Feigenbaum (1978, 1979) and Jonker & Rand (1981a, 1981b). Here we examine in detail one such function studied by Grassberger (1981), which, though apparently rather contrived, is easy to analyse and is of importance in a more general context.

Consider the functional equation, known as the renormalization equation,

$$g(g(x/\alpha)) = - g(x)/\alpha, \qquad (8.46)$$

together with the normalization condition that $g(0) = 1$. It may be shown with considerable difficulty (see Campanino & Epstein (1981) and Lanford (1982)) that this equation is satisfied by an even, real-analytic function g on

Fig. 8.5

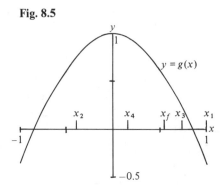

$[-1, 1]$ for a value of $\alpha = 2.50290\ldots$ (Whilst this solution is locally unique, it is not known if (8.46) has a solution for any other values of α.) The function g is given to four decimal places on $[-1, 1]$ by

$$g(x) = 1 - 1.5276\,x^2 + 0.1048\,x^4 + 0.0267\,x^6$$
$$- 0.0035\,x^8 + 0.0008\,x^{10} + 0.0003\,x^{12} \tag{8.47}$$

and its graph is shown in Figure 8.5. We use ideas from Section 8.3 to estimate the dimension of an attractor of g.

Theorem 8.21
The function g has a unique (unstable) periodic orbit of period 2^k for each $k \geq 0$, and a closed invariant attractor E with $0.5345 < \dim E < 0.5544$. If $x \in [-1, 1]$, then either $g^m(x)$ is a periodic point for m sufficiently large (this happens only for countably many x) or else the distance from $g^m(x)$ to E tends to 0 as $m \to \infty$. If $x \in E$ the orbit $\{g^m(x)\}_1^\infty$ is dense in E.

Proof. Iterating (8.46) and using the evenness of g

$$g^{2^k}\left(\left(-\frac{1}{\alpha}\right)^k x\right) = \left(-\frac{1}{\alpha}\right)^k g(x) \quad (k \geq 0). \tag{8.48}$$

Writing $x_m = g^m(0)$ we see that $x_{2^k} = (-1/\alpha)^k$ $(k \geq 0)$, and by calculation that

$$x_1 = 1, \quad x_2 = -0.3995, \quad x_3 = 0.7589, \quad x_4 = 0.1596.$$

Let V be the open interval (x_2, x_1) and define mappings ψ_1, ψ_2 on $\bar{V}$ by

$$\psi_1(x) = -\frac{1}{\alpha}x, \quad \psi_2(x) = g^{-1}\left(-\frac{1}{\alpha}x\right), \tag{8.49}$$

where g^{-1} is always taken to mean the positive value. We note that

$$\psi_1(V) = (x_2, x_4), \quad \psi_2(V) = (x_3, x_1)$$

so the open set condition holds for the contractions ψ_1 and ψ_2 (see Section

8.3). By Theorems 8.3 and 8.8 there is a unique compact set E such that $E = \psi_1(E) \cup \psi_2(E)$, with $s \leq \dim E \leq t$, where

$$(\inf |\psi_1'(x)|)^s + (\inf |\psi_2'(x)|)^s = 1 = (\sup |\psi_1'(x)|)^t + (\sup |\psi_2'(x)|)^t$$

with the infs. and sups. over $x \in \overline{V}$. Since the extreme values of g', and thus of $(g^{-1})'$, are attained at the ends of the interval $\overline{V}$ we get

$$\alpha^{-s} + (\alpha |g'(x_1)|)^{-s} = 1 = \alpha^{-t} + (\alpha |g'(x_3)|)^{-t},$$

so, calculating the derivatives, $0.5345 < \dim E < 0.5544$.

It remains to examine the limit behaviour of the iterates of the points. Since $g[-1, 1] \subset \overline{V}$ we may work entirely in $\overline{V}$. Rewriting (8.46) we see that $g(-(1/\alpha)x) = g(-(1/\alpha)g(x))$ (noting that $g(-(1/\alpha)x) > 0$), resulting in the functional identities

$$g \circ \psi_1 = \psi_2 \circ g, \quad g \circ \psi_2 = \psi_1 \tag{8.50}$$

valid on $[-1, 1]$. Thus writing $\psi_{j_1 j_2 \ldots j_k}$ for $\psi_{j_1} \circ \psi_{j_2} \circ \ldots \circ \psi_{j_k}$ we have

$$g \circ \psi_{11 \ldots 1} = \psi_{22 \ldots 2} \circ g,$$
$$g \circ \psi_{11 \ldots 12 j_1 \ldots j_k} = \psi_{22 \ldots 21 j_i \ldots j_k} \quad (2 \leq i \leq k+1). \tag{8.51}$$

Since $g(\overline{V}) \subset \overline{V}$ it follows that

$$g(\psi^k(\overline{V})) \subset \psi^k(\overline{V}), \tag{8.52}$$

where, as in Section 8.3, $\psi^k(\overline{V}) = \bigcup_{j_1 \ldots j_k} \psi_{j_1 \ldots j_k}(\overline{V})$. Thus any point in $\psi^k(\overline{V})$ is trapped in $\psi^k(\overline{V})$ under iterates of g. Further, we see from (8.51) that if $x \in \psi^k(\overline{V})$, then $g^m(x)$ visits each of the 2^k (disjoint) sets $\psi_{j_1 \ldots j_k}(\overline{V})$ in turn as m increases.

Let $x_f = 0.5493 \ldots$ be the unique fixed point of g in $[-1, 1]$. By (8.48) the points $(-1/\alpha)^k x_f$ and thus $g^m((-1/\alpha)^k x_f)$ are periodic with period, at most, 2^k. However, by the above remark $g^m((-1/\alpha)^k x_f)$ lies in a different $\psi_{j_1 \ldots j_k}(\overline{V})$ for every 2^k consecutive values of m (note that $(-1/\alpha)^k x_f \in \psi_{11 \ldots 1}(\overline{V})$), giving an orbit of period 2^k.

Now let $x \in \overline{V}$ be a point such that $g^m(x)$ never hits one of these periodic orbits. We show that $g^m(x)$ must converge to E. We already know that once inside $\psi^k(\overline{V})$ the iterates $g^m(x)$ are trapped there (8.52). We show that under iteration the points of $\psi^k(\overline{V})$ eventually pass into $\psi^{k+1}(\overline{V})$ for each k. Suppose that $g^m(x) \in \psi^k(\overline{V})$ for some $k \geq 0$, say that $g^m(x) = \psi_{j_1 \ldots j_k}(y)$, where $y \in \overline{V}$. The point y cannot be the fixed point x_f, otherwise $g^m(x)$ would have period 2^k, by (8.51). Thus either $y \in [x_2, x_4] \cup [x_3, x_1]$ or else $y \in [x_4, x_f) \cup (x_f, x_3]$. In the latter case, since $g'(x) < -1$ on $g^{-1}[x_4, x_3] = [-x_2, x_3]$, the iterates of y move away from the unstable fixed point x_f until $g^r(y) \in [x_2, x_4] \cup [x_3, x_1]$ for some r. Hence in either case we may find $r \geq 0$ such that $g^r(y) \in \psi_j(\overline{V})$ for $j = 1$ or 2. Iterating (8.51),

$g^{r2^k} \circ \psi_{j_1 \ldots j_k} = \psi_{j_1 \ldots j_k} \circ g^r$, so we conclude that

$$g^{r2^k + m}(x) = g^{r2^k}(\psi_{j_1 \ldots j_k}(y))$$
$$= \psi_{j_1 \ldots j_k}(g^r(y)) \subset \psi_{j_1 \ldots j_k}(\overline{V}) \subset \psi^{k+1}(\overline{V}).$$

Hence any point x of $\overline{V}$ that does not end up in a periodic orbit eventually enters $\psi^k(\overline{V})$ for all k. By (8.20) $E = \bigcap_{k=0}^{\infty} \psi^k(\overline{V})$, so the distance from $g^m(x)$ to E must tend to zero.

If $x \in E$, then for any k the iterates $g^m(x)$ visit each of the 2^k sets $\psi_{j_1 \ldots j_k}(\overline{V})$ in turn, from which it follows that the orbit $\{g^m(x)\}$ is dense in E.   □

Note that $\dim E$ may be found to any desired degree of accuracy by estimating the derivatives of $\psi_{j_1 \ldots j_k}$ for large enough k. In fact, $\dim E = 0.538\ldots$. Ruelle (1983) has recently shown that the attractors of such mappings are actually s-sets, where s is a real number which may be specified in terms of the 'pressure' of the mapping.

The mapping g described above is, in a sense, characteristic of a much larger class of mappings of $[-1, 1]$ which exhibit chaotic behaviour. Let f be any smooth (say twice differentiable) mapping of $[-1, 1]$ into itself such that f has a unique maximum at 0, with $f(0) = 1, f''(0) < 0, f'(x) > 0$ for $x < 0$ and $f'(x) < 0$ for $x > 0$. Let f_λ be the mapping of $[-1, 1]$ given by $f_\lambda(x) = \lambda f(x)$, where $0 < \lambda < 1$. Then for small λ the mapping f_λ has a single stable fixed point (stable in the sense that nearby points converge to it under iterations of f_λ). On increasing λ it may be shown that on reaching a value λ_1 this fixed point becomes unstable and gives way to a stable pair of points of period 2. Increasing λ further to λ_2 this cycle bifurcates into a stable cycle of period 4, at λ_3 a cycle of period 8 appears and so on. (See May (1976), Hofstadter (1981) or Feigenbaum (1981) for a more detailed description of this process, with particular reference to the 'logistic mapping' $f(x) = 1 - 2x^2$, equivalent to the mapping $f(x) = 4x(1 - x)$ on $[0, 1]$ after a coordinate transformation.) Feigenbaum (1978, 1979, 1981) has discovered some remarkable properties of the sequence $\{\lambda_j\}$; he shows that λ_j increases to a critical value λ_∞ as $j \to \infty$ in such a way that $(\lambda_{j+1} - \lambda_j)/(\lambda_{j+2} - \lambda_{j+1}) \to \delta$, where $\delta = 4.669\ldots$ is a universal constant, that is, does not depend on the exact form of the function f. The behaviour of the iterates of f_{λ_∞} is known to be qualitatively similar to that of the function g studied above: f_{λ_∞} always has a single unstable orbit of period 2^k for each k and an invariant set of Cantor-like character that attracts almost all points of the interval. On letting λ increase beyond λ_∞ stable periodic orbits again prevail, and these bifurcate in a similar way as λ approaches a second critical value corresponding to the accumulation of cycles of period $3 \cdot 2^j$, and so on.

Feigenbaum shows, by investigating the periodic points of f as $j \to \infty$, that for any function f of the type under consideration,

$$\lim_{k \to \infty} (-\alpha)^k f_{\lambda_\infty}^{2^k}(x/\alpha^k) = g(x)$$

satisfies the functional equation (8.46), where $\alpha = 2.5029\ldots$ is the number for which a solution of the equation is known to exist. (A linear scaling of coordinates allows the assumption $g(0) = 1$.) Thus as far as the limit behaviour of $f_{\lambda_\infty}^m$ as $m \to \infty$ is concerned, the iterates of f_{λ_∞} might be expected to behave in a similar manner to those of g, not only in a qualitative manner but also, in some ways, in a quantitative manner. In particular it might be hoped that the Hausdorff dimension of the attractor of f_{λ_∞} is always the same as that of g. Grassberger (1981) presents some computational evidence that this is indeed the case, that at the first accumulation of bifurcation points λ_∞, the strange attractor known to be present has dimension $0.538\ldots$.

Of course, similar problems occur in higher dimensions, and the analysis becomes considerably harder. A transformation of $\mathbb{R}^2$ due to Smale (1967), known as the 'horseshoe diffeomorphism' illustrates how strange attractors can occur as a result of 'folding and contraction'. The mapping f on the unit square S is defined to be qualitatively as indicated in Figure 8.6(a), where $f(a) = a'$, etc., with the upper and lower halves of the square mapped onto the two arms of the horseshoe. The first few iterates of $f^m(S)$ are indicated in Figure 8.6(b) and it is clear that the iterates of all points of S converge to the Cantor-like set $\bigcap_{m=0}^{\infty} f^m(S)$. By defining f in a suitable manner it is

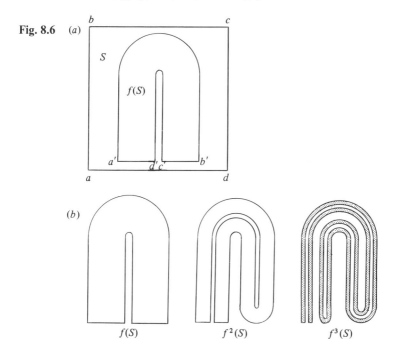

Fig. 8.6 (a)

(b)

$f(S)$ $f^2(S)$ $f^3(S)$

possible to obtain an attractor of any dimension between 1 and 2. Because of the width-contracting effect of f the attractor is stable under perturbations of f.

A mapping of $\mathbb{R}^2$ that is harder to analyse is the Hénon–Pomeau mapping given in coordinate form by

$$f(x, y) = (y + 1 - ax^2, bx),$$

where a and b are constants. This mapping has a Jacobian of $-b$ at all (x, y), so contracts areas at a uniform rate. It has attracted much attention since, to within a linear change of coordinates, it is the most general quadratic mapping of $\mathbb{R}^2$ with this property. It is related to the 1-dimensional (quadratic) logistic mapping and its attractor might be expected to have a fine structure for certain (a, b). However, for values of a and b which are known to yield an invariant set of fractional dimension, it turns out that this set is not an attractor. Hénon & Pomeau (1976) investigate some of the orbits for $a = 1.4$ and $b = 0.3$. For a more recent analysis see Simó (1979) and also the article by Whitley (1983).

We have so far considered only discrete dynamical systems, but very similar ideas hold for continuous systems or *flows*, that is, for solution curves of systems of differential equations. Write $f_t(x)$ for the position reached by the point x at time t. Then

$$f_{t_2}(f_{t_1}(x)) = f_{t_1 + t_2}(x),$$

assuming that the flow is autonomous (that is, derived from time-independent differential equations). Again, one seeks closed invariant sets (for which $f_t(E) \subset E$ for all $t \geq 0$) that are attractors in the sense that $f_1(x)$ visits every neighbourhood of any point of E for arbitrarily large t, for a large set of x.

A simple modification of Smale's horseshoe mapping provides one example of how strange attractors can occur for flows. Let f_t be a flow inside a ring V of square section that proceeds round the inside of the ring at a constant rate, taking unit time to complete a circuit of the ring. Let Π be a plane transverse to the ring and suppose that f_t is defined so that $f_1(x) = f(x)$ if $x \in S = V \cap \Pi$, where f is the horseshoe mapping on S. (It is straightforward to define a flow on V with this property.) Then if E is the attractor for the flow, $E \cap \Pi = \bigcap_{m=0}^{\infty} f^m(S)$ is the attractor for the horseshoe mapping. To find the remainder of the attractor we simply follow this set around the ring to get $E = \bigcup_{t=0}^{1} f_t(\bigcap_{m=0}^{\infty} f^m(S))$. Thus the attractor is locally a product of the horseshoe attractor and a line segment, so has, by a minor variation of the work of Section 5.3, dimension equal to $1 + \dim \bigcap_0^{\infty} f^m(S)$. Alternatively, the attractor may be thought of locally as the product of a 1-dimensional Cantor set and a plane surface.

In practice, the dynamics of flows of differential equations can be exceedingly complex and this sort of simplistic approach does not carry very far. Perhaps the system of differential equations most examined in the search for strange attractors are the equations proposed by Lorenz (1963) to model aspects of the weather

$$\left.\begin{aligned} \dot{x} &= \sigma(y - x), \\ \dot{y} &= rx - y - xz, \\ \dot{z} &= xy - bz, \end{aligned}\right\} \tag{8.53}$$

where σ, r and b are constants. A thorough account of the attractors and bifurcations of these equations is given in the book by Sparrow (1982). Chaotic behaviour occurs for certain values of the parameters, $\sigma = 10, r = 28, b = \frac{8}{3}$ being the case usually selected for study.

Temam (1983) examines the Navier–Stokes equation of fluid dynamics. In particular, he estimates the Hausdorff dimension of an attractor, as well as the dimension of the singularities of weak solutions as a subset of space-time.

Recently, attempts have been made to express the dimension of attractors of dynamical systems in terms of constants of the motion, in particular in terms of the Lyapunov characteristic exponents of the orbits. (The Lyapunov exponents are the long-term orbit averages of the eigenvalues of the localized expansion matrices.) For a survey of these ideas, together with some examples, see Frederickson, Kaplan, Yorke & Yorke (1983).

For further discussion of attractors and their relationship with problems of turbulence, etc., see the volumes of conference proceedings: Temam (1976), Bernard & Raţiu (1977), Haken (1982), Guckenheimer & Holmes (1983) and Barenblatt, Iooss & Joseph (1983). A vast amount of work remains to be done on attractors, indeed many of the results that have been claimed have still to be put on a rigorous foundation.

8.8 Brownian motion

In 1827 the botanist R. Brown observed that minute particles suspended in a liquid were in constant motion and described highly irregular paths. This was later explained as resulting from molecular bombardment of the particle. A similar phenomenon was noticed in smoke particles in air. In 1905 Einstein published a mathematical study of this motion, which eventually led to Perrin's Nobel prize-winning calculation of Avogardro's number.

A rigorous probabilistic model of Brownian motion was proposed by Wiener (1923). He constructed the 'Wiener process' which exhibits random

behaviour very similar to that of Brownian motion. Here we outline the
details of this model and give Taylor's (1953) proof that Brownian paths
have Hausdorff dimension 2 with probability one.

If $n \geq 1$, let Ω denote the class of all continuous paths $\omega:[0, \infty) \rightarrow \mathbb{R}^n$
which have $\omega(0)$ as the origin. We think of $\omega(t)$ as the position at time t of a
particle describing the path ω. It may be shown that there exists a
probability measure p (with $p(\Omega) = 1$) defined on a large system of subsets of
Ω such that:

(a) the paths have *independent increments*, that is, $\omega(t_2) - \omega(t_1)$ and
$\omega(t_4) - \omega(t_3)$ are independent if $t_1 \leq t_2 \leq t_3 \leq t_4$

(b) $\omega(t + h) - \omega(t)$ has Gaussian distribution with zero mean and variance h
for all t. In particular, the distribution of the path increments is *stationary*
($\omega(t + h) - \omega(t)$ does not depend on t), is *isotropic* (independent of
direction), and is such that

$$p\{\omega : |\omega(t + h) - \omega(t)| \leq \rho\}$$
$$= ch^{-n/2} \int_0^\rho r^{n-1} \exp(-r^2/2h) dr, \qquad (8.54)$$

where c is a normalization constant chosen to ensure that $p(\Omega) = 1$.

The probability measure p may be constructed so that a large class
of sets of paths are measurable, including all subsets of Ω of the form
$\{\omega : \omega(t_i) \in B_i (i = 1, 2, \ldots, k)\}$, where the B_i are Borel subsets of $\mathbb{R}^n$. It is quite
complicated to establish the existence and uniqueness of such a process, the
interested reader is referred, for example, to Hida (1980). In fact, the
measure p may be defined on the class of all (not necessarily continuous)
functions $\omega:[0, \infty) \rightarrow \mathbb{R}^n$, and it follows from the other conditions that the
paths are continuous with probability one.

It is easy to see that projecting a Brownian process in $\mathbb{R}^n$ onto a subspace
gives a lower-dimensional Brownian process.

The importance of the Brownian probability distribution is that it is the
essentially unique distribution for which the paths have stationary,
independent, isotropic increments of finite variance. These are conditions
which are likely to apply in any physical situation.

Brownian motion may be thought of as the limiting case of a random
walk as the step length tends to zero. The process is *statistically self-similar*
in the sense that the paths $\omega(t)$ and $\omega(\gamma^2 t)/\gamma$ have the same probability
distribution for any $\gamma > 0$.

Lemma 8.22
If $0 < \lambda < \frac{1}{2}$, then for almost all $\omega \in \Omega$ there exists $h_0 > 0$ such that
$$|\omega(t + h) - \omega(t)| \leq |h|^\lambda$$
for $0 \leq t \leq 1$ and $|h| < h_0$.

Proof. Let j and m be positive integers. Then (8.54) implies that

$$p\{\omega:|\omega(2^{-j}m) - \omega(2^{-j}(m-1))| > 2^{-j\lambda}\}$$

$$= c2^{jn/2} \int_{2^{-j\lambda}}^{\infty} r^{n-1} \exp(-r^2 2^{j-1})dr$$

$$= c \int_{2^{j(\frac{1}{2}-\lambda)}}^{\infty} r^{n-1} \exp(-r^2/2)dr$$

$$\leq c_1 \exp(-2^{j(\frac{1}{2}-\lambda)})$$

$$\leq c_2 2^{-2j}$$

after a substitution and some sweeping estimates valid if $\lambda < \frac{1}{2}$, where c_1 and c_2 are independent of j and m. Hence,

$$p\{\omega:|\omega(2^{-j}m) - \omega(2^{-j}(m-1))| > 2^{-j\lambda}$$

$$\text{for some } j,m \text{ with } j > k \text{ and } 1 \leq m \leq 2^j\}$$

$$\leq c_2 \sum_{j=k}^{\infty} 2^j 2^{-2j} = c_2 2^{-k}.$$

Thus, for almost all $\omega \in \Omega$,

$$|\omega(2^{-j}m) - \omega(2^{-j}(m-1))| \leq 2^{-j\lambda} \quad \text{for} \quad m = 1, 2, \ldots, 2^j \qquad (8.55)$$

for all j sufficiently large. We may express any interval $(t, t+h) \subset (0,1)$ as a countable union of contiguous binary intervals of the form $(2^{-j}(m-1), 2^{-j}m)$ with $2^{-j} \leq h$ and with, at most, two intervals of each length. (Take all binary subintervals of $(t, t+h)$ not contained in any other such intervals.) Then if h is small enough and k is the least integer with $2^{-k} \leq h$.

$$|\omega(t+h) - \omega(t)| \leq 2 \sum_{j=k}^{\infty} 2^{-j\lambda} = 2^{-k\lambda} 2/(1 - 2^{-\lambda}) \leq h^{\lambda} 2/(1 - 2^{-\lambda}),$$

using (8.55) and the triangle inequality, together with the continuity of $\omega(t)$. This is true for $0 \leq t < t + h \leq 1$ if h is sufficiently small, for almost all $\omega \in \Omega$ for any $\lambda < \frac{1}{2}$, so the conclusion of the lemma follows. $\square$

Theorem 8.23
Brownian paths in $\mathbb{R}^n (n \geq 2)$ have Hausdorff dimension 2 with probability one.

Proof. (a) We use a potential-theoretic method to show that $\dim \omega \geq 2$ for almost all paths ω. Fix $1 < s < 2$. Using (8.54) we get

$$\int_{\omega \in \Omega} \frac{dp(\omega)}{|\omega(t+h) - \omega(t)|^s}$$

$$= ch^{-n/2} \int_0^{\infty} r^{-s+n-1} \exp(-r^2/2h)dr = c_1 h^{-s/2}$$

on substituting for r^2/h in the integrand, where c_1 is independent of h and t.

Thus

$$\int_{u=0}^{1}\int_{t=0}^{1}\int_{\omega\in\Omega}\frac{dp(\omega)dtdu}{|\omega(t)-\omega(u)|^s}=\int_0^1\int_0^1\frac{c_1 dtdu}{|t-u|^{s/2}}<\infty.$$

Then measurability properties of Brownian motion ensure that $|\omega(t)-\omega(u)|^{-s}$ is measurable with respect to the product measure on $[0,1]\times[0,1]\times\Omega$, so we conclude, using Fubini's theorem, that

$$\int_0^1\int_0^1\frac{dtdu}{|\omega(t)-\omega(u)|^s}<\infty$$

for almost all $\omega\in\Omega$.

There is a natural mass distribution μ_ω on the path ω, given by $\mu_\omega(E)=\mathscr{L}^1\{t:0\le t\le 1 \text{ and } \omega(t)\in E\}$. Thus for almost all $\omega\in\Omega$ the s-energy $I_s(\mu_\omega)$ is finite, so, by Corollary 6.6(a), $s\le\dim(\omega[0,1])\le\dim\omega$. This is true for all $s<2$, so $\dim\omega\ge 2$ for almost all $\omega\in\Omega$.

(b) We use a variant of the method of Theorem 8.1 to obtain an upper bound for the dimension of the paths. Take $s>2$. Lemma 8.22 implies that for almost all $\omega\in\Omega$ we may find $h_0>0$ such that

$$|\omega(t+h)-\omega(t)|\le|h|^{1/s} \quad \text{if} \quad |h|\le h_0 \quad \text{and} \quad 0\le t<t+h\le 1.$$

For such a path ω the subarc $\omega[t,t+h]$ may be enclosed in a ball of radius $h^{1/s}$ if $h\le h_0$. Consequently, if m is an integer with $1/m\le h_0$, we may enclose $\omega[0,1]=\bigcup_{j=1}^m\omega[(j-1)/m,j/m]$ in m balls, each of radius $m^{-1/s}$. Thus $\mathscr{H}_\delta^s(\omega[0,1])\le m2^s(m^{-1/s})^s=2^s$ if $\delta\ge 2m^{-1/s}$. Letting $m\to\infty$ we conclude that $\mathscr{H}^s(\omega[0,1])\le 2^s$ if $s>2$, so in fact, by (1.14), $\mathscr{H}^s(\omega[0,1])=0$ and thus $\mathscr{H}^s(\omega)=0$. □

Taylor (1953) also shows that $\mathscr{H}^s(\omega)=0$ for almost all paths in $\mathbb{R}^n$. More delicate calculations carried out by Ray (1963) and Taylor (1964) show that with probability one the Brownian paths have positive but finite Hausdorff measure with respect to the measure function

$$h(t)=t^2\log\frac{1}{t}\log\log\log\frac{1}{t}$$

(In the definition of Hausdorff measure $|U_i|^s$ is replaced by $h(U_i)$ and this allows a finer definition of dimension, see Section 1.2.)

Similar questions have been asked about the almost sure Hausdorff dimension of double, triple and multiple points of Brownian paths. The theory has also been extended to more general Lévy stable processes. See Rogers (1970, p. 133), Pruitt & Taylor (1969), Taylor (1973), Pruitt (1975), Adler (1981) and Mandelbrot (1982, Section 39) for further discussion of some of these topics and for numerous further references.

8.9 Conclusion

In this chapter we have given accounts of a few of the situations in which fractal sets occur and to which the ideas discussed earlier in the book may be applied. Mandelbrot's (1975, 1977, 1982) essays provide a vast panorama of other such sets, many of which have not yet been subjected to rigorous mathematical analysis. These essays contain very complete references to further topics, as does the final chapter of Roger's (1970) book on applications of Hausdorff measure.

Although much of the work that has been described in this book is not particularly recent, the subject contains many unsolved problems, some old and some new. The theory and applications of the geometry of fractal sets promises to be an active area of research for many years to come. In words from Besicovitch's film on the Kakeya problem, 'may the efforts of those who research into these matters be crowned with success'.

Exercises on Chapter 8

8.1   Show that Theorem 8.1 remains true if c in (8.1) is allowed to depend on x.

8.2   Prove that the graph of a continuously differentiable function on $\mathbb{R}$ is a 1-set.

8.3   Show that the graph of the Weierstrass function (8.3) has dimension, at most, s.

8.4   Let $f : [0, 1] \to \mathbb{R}$ have graph Γ. Show that if $\sum_{i=1}^{m} |f(x_i) - f(x_{i-1})|^s \le c$ for all dissections $0 \le x_0 < \cdots < x_m \le 1$, then $\mathcal{H}^s(\Gamma) < \infty$.

8.5   Suppose that the open set condition (8.19) holds for the contractions $\{\psi_j\}_1^m$ on $\mathbb{R}^n$. Suppose further that for each j and all x, y

$$q_j |x - y| \le |\psi_j(x) - \psi_j(y)| \le r_j |x - y|.$$

If E is the invariant set associated with these contractions adapt the proof of Theorem 8.6 to show that $s' \le \dim E \le t$, where $s' = s - \log (\max (r_j/q_j))/\log (\min (1/q_j))$ and where $\sum_1^m q_j^s = 1 = \sum_1^m r_j^t$.

8.6   Let $\{B_i\}_1^\infty$ be an osculatory packing of an open set $V \subset \mathbb{R}^n$. Show that $V \subset \bigcup_i B_i'$, where B_i' is the ball concentric with B_i but with twice the radius. Deduce that the Hausdorff dimension of the residual set is, at most, equal to the exponent of the packing.

8.7   Show that the residual set of the packing of equilateral triangles shown in Figure 8.4 is an s-set for $s = \log 3/\log 2$. (Note that the residual set is self-similar.)

8.8   Show that the mapping $f : \mathbb{R} \to \mathbb{R}$ given by $f(x) = \frac{3}{2}(1 - |2x - 1|)$ has 0 as a fixed point and the Cantor set E as an invariant set. Show that if $x \in E \setminus \{0\}$, then the iterates $f^m(x)$ are dense in E, and that if $x \notin E$, then $f^m(x) \to -\infty$ as $m \to \infty$. (You may find it helpful to consider base three expansions.)

References

The numbers in square brackets indicate the sections in which each reference is cited.

Adler, R.J. (1981). *The Geometry of Random Fields.* New York: Wiley. [8.8]

Alexander, R. (1975). 'Random compact sets related to the Kakeya problem'. *Proceedings of the American Mathematical Society*, **53**, 415–19. [7.2]

van Alphén, H.J. (1942). 'Uitbreiding van een stelling von Besicovitch'. *Mathematica, Zutphen, B*, **10**, 144–57. [7.2]

Anderson, R.D. & Klee, V.L. (1952). 'Convex functions and upper semi-continuous functions'. *Duke Mathematical Journal*, **19**, 349–57. [8.6]

Baker, A. (1975). *Transcendental Number Theory.* Cambridge University Press. [8.5]

Baker, A. & Schmidt, W.M. (1970). 'Diophantine approximation and Hausdorff dimension'. *Proceedings of the London Mathematical Society* (3), **21**, 1–11. [8.5]

Baker, R.C. (1978). 'Dirichlet's theorem on Diophantine approximation'. *Mathematical Proceedings of the Cambridge Philosophical Society*, **83**, 37–59. [8.5]

Barenblatt, G. I., Iooss, G & Joseph, D. D. (eds.). (1983). *Nonlinear Dynamics and Turbulence.* London: Pitman.

Beardon, A.F. (1965). 'On the Hausdorff dimension of general Cantor sets'. *Proceedings of the Cambridge Philosophical Society*, **61**, 679–94. [1.5]

Bernard, P. & Ratiu, T. (eds.). (1977). Turbulence Seminar (Berkley). *Lecture Notes in Mathematics* **615**, New York: Springer. [8.7]

Berry, M.V. & Lewis, Z.V. (1980). 'On the Weierstrass–Mandelbrot fractal function'. *Proceedings of the Royal Society of London*, **A370**, 459–84. [8.2]

Besicovitch, A.S. (1919). 'Sur deux questions d'intégrabilité des fonctions. *J. Soc. Phys.-Math. (Perm')*, **2**, 105–23. [7.1, 7.2]

Besicovitch, A.S. (1928a). 'On the fundamental geometrical properties of linearly measurable plane sets of points'. *Mathematische Annalen*, **98**, 422–64. [1.2, 2.2, 3.1, 3.3, 3.4, 3.5, 6.4]

Besicovitch, A.S. (1928b). 'On Kakeya's problem and a similar one'. *Mathematische Zeitschrift*, **27**, 312–20. [7.1, 7.2]

Besicovitch, A.S. (1929). 'On linear sets of points of fractional dimension'. *Mathematische Annalen*, **101**, 161–93. [2.2]

Besicovitch, A.S. (1934a). 'On tangents to general sets of points'. *Fundamenta Mathematicae*, **22**, 49–53. [3.2]

Besicovitch, A.S. (1934b). 'Sets of fractional dimensions IV: On rational approximation to real numbers'. *Journal of the London Mathematical Society*, **9**, 126–31. [8.5]

Besicovitch, A.S. (1938). 'On the fundamental geometric properties of linearly measurable plane sets of points ɪɪ'. *Mathematische Annalen*, **115**, 296–329. [1.2, 2.2, 3.1, 3.2, 3.3, 3.4, 3.5]

Besicovitch, A.S. (1939). 'On the fundamental geometric properties of linearly measurable plane sets of points ɪɪɪ'. *Mathematische Annalen*, **116**, 349–57. [6.2, 6.4]

Besicovitch, A.S. (1942). 'A theorem on s-dimensional measure of sets of points'. *Proceedings of the Cambridge Philosophical Society*, **38**, 24–7. [1.2]

Besicovitch, A.S. (1944). 'On the existence of tangents to rectifiable curves'. *Journal of the London Mathematical Society*, **19**, 205–7. [3.2]

Besicovitch, A.S. (1945a). 'A general form of the covering principle and relative differentiation of additive functions'. *Proceedings of the Cambridge Philosophical Society*, **41**, 103–10. [1.3, 2.2]

Besicovitch, A.S. (1945*b*). 'On the definition and value of the area of a surface'. *Quarterly Journal of Mathematics (Oxford)*, **16**, 88–102. [3.5]

Besicovitch, A.S. (1946). 'A general form of the covering principle and relative differentiation of additive functions II'. *Proceedings of the Cambridge Philosophical Society*, **42**, 1–10. [1.3]

Besicovitch, A.S. (1947). 'Corrections to Besicovitch (1946)'. *Proceedings of the Cambridge Philosophical Society*, **43**, 590. [1.3]

Besicovitch, A.S. (1948). 'On distance sets'. *Journal of the London Mathematical Society*, **23**, 9–14. [7.4]

Besicovitch, A.S. (1949). 'Parametric surfaces II. Lower semi-continuity of the area'. *Proceedings of the Cambridge Philosophical Society*, **45**, 14–23. [3.5]

Besicovitch, A.S. (1950). 'Parametric surfaces'. *Bulletin of the American Mathematical Society*, **56**, 288–96. [3.5]

Besicovitch, A.S. (1952). 'On existence of subsets of finite measure of sets of infinite measure'. *Indagationes Mathematicae*, **14**, 339–44. [5.1, 5.2]

Besicovitch, A.S. (1954). 'On approximation in measure to Borel sets by F_σ-sets'. *Journal of the London Mathematical Society*, **29**, 382–3. [1.2]

Besicovitch, A.S. (1956). 'On the definition of tangents to sets of infinite linear measure'. *Proceedings of the Cambridge Philosophical Society*, **52**, 20–9. [3.2]

Besicovitch, A.S. (1957). 'Analysis of tangential properties of curves of infinite length'. *Proceedings of the Cambridge Philosophical Society*, **53**, 69–72. [3.2]

Besicovitch, A.S. (1960). 'Tangential properties of sets and arcs of infinite linear measure'. *Bulletin of the American Mathematical Society*, **66**, 353–9. [3.2]

Besicovitch, A.S. (1963*a*). 'On singular points of convex surfaces'. In *Proceedings of Symposia in Pure Mathematics*, 7, 21–3. Providence: American Mathematical Society. [8.6]

Besicovitch, A.S. (1963*b*). 'On the set of directions of linear segments on a convex surface'. In *Proceedings of Symposia in Pure Mathematics*, 7, 24–5. Providence: American Mathematical Society. [8.6]

Besicovitch, A.S. (1963*c*). 'The Kakeya problem'. *American Mathematical Monthly*, **70**, 697–706. [7.2]

Besicovitch, A.S. (1964*a*). 'On fundamental geometric properties of plane line-sets'. *Journal of the London Mathematical Society*, **39**, 441–8. [7.1, 7.3]

Besicovitch, A.S. (1964*b*). 'On one-sided densities of arcs of positive two-dimensional measure'. *Proceedings of the Cambridge Philosophical Society*, **60**, 517–24. [8.2]

Besicovitch, A.S. (1965*a*). 'Arcs and chords'. *Proceedings of the London Mathematical Society* (3), **14A**, 28–37. [8.2]

Besicovitch, A.S. (1965*b*). 'On arcs that cannot be covered by an open equilateral triangle of side 1'. *Mathematical Gazette*, **49**, 286–8. [7.4]

Besicovitch, A.S. (1967). 'Arcs and chords II'. *Journal of the London Mathematical Society*, **42**, 86–92. [8.2]

Besicovitch, A.S. (1968). 'On linear sets of points of fractional dimensions II'. *Journal of the London Mathematical Society*, **43**, 548–50. [4.2]

Besicovitch, A.S. & Miller, D.S. (1948). 'On the set of distances between the points of a Carathéodory linearly measurable plane point set'. *Proceedings of the London Mathematical Society* (2), **50**, 305–16. [7.4]

Besicovitch, A.S. & Moran, P.A.P. (1945). 'The measure of product and cylinder sets'. *Journal of the London Mathematical Society*, **20**, 110–20. [5.1, 5.3]

Besicovitch, A.S. & Rado, R. (1968). 'A plane set of measure zero containing circumferences of every radius'. *Journal of the London Mathematical Society*, **43**, 717–9. [7.3]

Besicovitch, A.S. & Taylor, S.J. (1952). 'On the set of distances between points of a general metric space'. *Proceedings of the Cambridge Philosophical Society*, **48**, 209–14. [7.4]

Besicovitch, A.S. & Taylor, S.J. (1954). 'On the complementary intervals of linear closed sets of zero Lebesgue measure'. *Journal of the London Mathematical Society*, **29**, 449–59. [1.5]

Besicovitch, A.S. & Ursell, H.D. (1937). 'Sets of fractional dimensions, v: On dimensional numbers of some continuous curves'. *Journal of the London Mathematical Society*, **12**, 18–25. [8.2]

Besicovitch, A.S. & Walker, G. (1931). 'On the density of irregular linearly measurable sets of points'. *Proceedings of the London Mathematical Society* (2), **32**, 142–53. [3.3]

Best, E. (1942). 'On sets of fractional dimensions III', *Proceedings of the London Mathematical Society* (2), **47**, 436–54. [1.5]

Blank, A.A. (1963). 'A remark on the Kakeya problem'. *American Mathematical Monthly*, **70**, 706–11. [7.4]

Blaschke, W. (1916). *Kreis und Kugel*. Leipzig. [3.2]

Borel, E. (1895). 'Sur quelques points de la théorie des fonctions'. *Ann. École Norm. sup.* (3), **12**, 9–55. [Int.]

Boyd, D.W. (1973a). 'Improved bounds for the disc packing constant'. *Aequationes Mathematicae*, **9**, 99–106. [8.4]

Boyd, D.W. (1973b). 'The residual set dimension of the Apollonian packing'. *Mathematika*, **20**, 170–4. [8.4]

Burkill, J.C. (1971). 'Abram Samoilovitch Besicovitch'. *Biographical Memoirs of Fellows of the Royal Society*, **17**, 1–16. [Int.]

Burkill, J.C. & Burkill, H. (1970). *A Second Course in Mathematical Analysis*. Cambridge University Press. [3.2]

Burton, G. (1979). 'The measure of the s-skeleton of a convex body'. *Mathematika*, **26**, 290–301. [8.6]

Campanino, M. & Epstein, H. (1981). 'On the existence of Feigenbaum's fixed point'. *Communications in Mathematical Physics*, **79**, 261–302. [8.7]

Carathéodory, C. (1914). 'Über das lineare Mass von Punktmengeneine Verallgemeinerung des Längenbegriffs'. *Nach. Ges. Wiss. Göttingen*, pp. 404–26. [Int.]

Carleson, L. (1967). *Selected Problems on Exceptional Sets*. Princeton: Van Nostrand. [6.2]

Casas, A. (1978). 'Aplicaciones de la teoría de medida lineal'. Tesis, Univ. Complutense de Madrid. [7.2]

Casas, A. & de Guzmán, M. (1981). 'On the existence of Nikodym sets and related topics'. *Rendiconti del Circolo Matematico de Palermo* (2), 1981, supp. No. 1, 69–73. [7.2]

Cigler, J. & Volkmann, B. (1963). 'Über die Häufigkeit von Zahlenfolgen mit gegebener Verteilungsfunktion'. *Abhandlungen aus dem Mathematischen Seminar der Universität Hamberg*, **26**, 39–54. [1.5]

Coxeter, H.S.M. (1961). *Introduction to Geometry*. New York: Wiley. [8.4]

Croft, H.T. (1965). Review of Besicovitch (1964). *Mathematical Reviews*, **30**, # 2122 [7.3]

Cunningham, F. (1971). 'The Kakeya problem for simply connected and for star-shaped sets'. *American Mathematical Monthly*, **78**, 114–29. [7.2, 7.4]

Cunningham, F. (1974). 'Three Kakeya problems'. *American Mathematical Monthly*, **81**, 582–92. [7.4]

Cunningham, F. & Schoenberg, I.J. (1965). 'On the Kakeya constant'. *Canadian Journal of Mathematics*, **17**, 946–56. [7.4]

Dalla, L. & Larman, D.G. (1980). 'Convex bodies with almost all k-dimensional sections polytopes'. *Mathematical Proceedings of the Cambridge Philosophical Society*, **88**, 395–401. [8.6]

Darst, R. & Goffman, C. (1970). 'A Borel set which contains no rectangle'. *American Mathematical Monthly*, **77**, 728–9. [7.4]

Davies, R.O. (1952a). 'On accessibility of plane sets and differentiation of functions of two real variables'. *Proceedings of the Cambridge Philosophical Society*, **48**, 215–32. [7.2, 7.4]

Davies, R.O. (1952*b*). 'Subsets of finite measure in analytic sets'. *Indagationes Mathematicae*, **14**, 488–9. [5.2]

Davies, R.O. (1956). 'A property of Hausdorff measure'. *Proceedings of the Cambridge Philosophical Society*, **52**, 30–4. [1.2]

Davies, R.O. (1965). 'A regular line-set covering finite measure'. *Journal of the London Mathematical Society*, **40**, 503–8. [7.3]

Davies, R.O. (1968). 'A theorem on the existence of non-σ-finite subsets'. *Mathematika*, **15**, 60–2. [5.2]

Davies, R.O. (1969). 'Measures of Hausdorff type'. *Journal of the London Mathematical Society* (2), **1**, 30–4. [1.2]

Davies, R.O. (1970). 'Increasing sequences of sets and Hausdorff measure'. *Proceedings of the London Mathematical Society* (3), **20**, 222–36. [5.1]

Davies, R.O. (1971). 'Some remarks on the Kakeya problem'. *Proceedings of the Cambridge Philosophical Society*, **69**, 417–21. [7.3, 7.4]

Davies, R.O. (1972). 'Another thin set of circles'. *Journal of the London Mathematical Society* (2), **5**, 191–2. [7.3]

Davies, R.O. (1979). 'Two counter-examples concerning Hausdorff dimensions of projections'. *Colloquium Mathematicum*, **42**, 53–8. [6.3]

Davies, R.O. (1982). 'Some counter-examples in measure theory'. *Proceedings of the Conference Topology and Measure* III. (Vitte, 1980), Part I, pp. 49–55. Greifswald: Ernst-Moritz-Arndt Universität. [6.3]

Davies, R.O. & Fast, H. (1978) 'Lebesgue density influences Hausdorff measure; large sets surface-like from many directions'. *Mathematika*, **25**, 116–19. [1.5]

Davies, R.O., Marstrand, J.M. & Taylor, S.J. (1960). 'On the intersection of transforms of linear sets'. *Colloquium Mathematicum*, **7**, 237–43. [7.4]

Davies, R.O. & Samuels, P. (1974). 'Density theorems for measures of Hausdorff type'. *Bulletin of the London Mathematical Society*, **6**, 31–6. [1.2, 2.2]

Dekking, F.M. (1982). 'Recurrent sets'. *Advances in Mathematics*, **44**, 78–104. [8.3]

Dickinson, D.R. (1939). 'Study of extreme cases with respect to the densities of irregular linearly measurable plane sets of points'. *Mathematische Annalen*, **116**, 359–73. [2.2]

Dunford, N. & Schwartz, J.T. (1958). *Linear Operators, Part I*. New York: Interscience. [3.2]

Eggleston, H.G. (1949). 'Note on certain *s*-dimensional sets'. *Fundamenta Mathematicae*, **36**, 40–3. [7.4]

Eggleston, H.G. (1950*a*). 'The Besicovitch dimension of Cartesian product sets'. *Proceedings of the Cambridge Philosophical Society*, **46**, 383–6. [5.3]

Eggleston, H.G. (1950*b*). 'A geometrical property of sets of fractional dimension'. *Quarterly Journal of Mathematics, Oxford Series* (2), **1**, 81–5. [5.3]

Eggleston, H.G. (1952). 'Sets of fractional dimensions which occur in some problems of number theory'. *Proceedings of the London Mathematical Society* (2), **54**, 42–93. [1.5, 8.5]

Eggleston, H.G. (1953*a*). 'On closest packing by equilateral triangles'. *Proceedings of the Cambridge Philosophical Society*, **49**, 26–30. [8.4]

Eggleston, H.G. (1953*b*). 'A correction to a paper on the dimension of Cartesian product sets'. *Proceedings of the Cambridge Philosophical Society*, **49**, 437–40. [5.3]

Eggleston, H.G. (1957). *Problems in Euclidean Space-Applications of Convexity*. London: Pergamon. [7.4]

Eggleston, H.G. (1958). *Convexity*. Cambridge University Press. [1.4]

Erdös, P. (1946), 'On the Hausdorff dimension of some sets in Euclidean space'. *Bulletin of the American Mathematical Society*, **52**, 107–9. [1.5]

Erdös, P. & Gillis, J. (1937). 'Note on the transfinite diameter'. *Journal of the London Mathematical Society*, **12**, 185–92. [6.2]

Ernst, L.R. & Freilich, G. (1976). 'A Hausdorff measure inequality'. *Transactions of the American Mathematical Society*, **219**, 361–8. [5.3]

Ewald, G., Larman, D.G. & Rogers, C.A. (1970). 'The directions of the line segments and of the r-dimensional balls in the boundary of a convex body in Euclidean space'. *Mathematika*, **17**, 1–20. [8.6]

Faber, V., Mycielski, J. & Pedersen, P. (1983). 'On the shortest curve which meets all the lines which meet a circle'. *Annales Polonici Mathematici*, to appear. [3.2]

Falconer, K.J. (1980a). 'Continuity properties of k-plane integrals and Besicovitch sets. *Mathematical Proceedings of the Cambridge Philosophical Society*, **87**, 221–6. [7.3]

Falconer, K.J. (1980b). 'Sections of sets of zero Lebesgue measure'. *Mathematika*, **27**, 90–6. [7.3]

Falconer, K.J. (1982). 'Hausdorff dimension and the exceptional set of projections'. *Mathematika*, **29**, 109–15. [6.3, 7.4]

Falconer, K.J. (1984). 'On a problem of Erdös on sequences and measurable sets'. *Proceedings of the American Mathematical Society*, **90**, 77–8. [7.4]

Federer, H. (1947). 'The (ϕ, k) rectifiable subsets of n-space'. *Transactions of the American Mathematical Society*, **62**, 114–92. [Int., 3.5, 6.4]

Federer, H. (1969). *Geometric Measure Theory*. New York: Springer. [Int., 3.5, 6.4, 6.5]

Fefferman, C. (1971). 'The multiplier problem for the ball'. *Annals of Mathematics*, **94**, 330–6. [7.5]

Feigenbaum, M.J. (1978). 'Quantitative universality for a class of non-linear transformations'. *Journal of Statistical Physics*, **19**, 25–52. [8.7]

Feigenbaum, M.J. (1979). 'The universal metric properties of nonlinear transformations'. *Journal of Statistical Physics*, **21**, 669–706. [8.7]

Feigenbaum, M.J. (1981). 'Universal behaviour in non-linear systems'. *Los Alamos Science*, **1**, 4–27. [8.7]

Fisher, B. (1973). 'On a problem of Besicovitch'. *American Mathematical Monthly*, **80**, 785–7. [7.2]

Frederickson, P., Kaplan, J., Yorke, E. & Yorke, J. (1983). 'The Lyapunov dimension of strange attractors'. *Journal of Differential Equations*, **49**, 185–207. [8.7]

Freilich, G. (1950). 'On the measure of Cartesian product sets'. *Transactions of the American Mathematical Society*, **69**, 232–75. [5.3]

Freilich, G. (1965). 'Carathéodory measure of cylinders'. *Transactions of the American Mathematical Society*, **114**, 384–400. [5.3]

Freilich, G. (1966). 'Gauges and their densities'. *Transactions of the American Mathematical Society*, **122**, 153–62. [2.2]

Frostman, O. (1935). 'Potential d'équilibre et capacité des ensembles avec quelques applications à la théorie des fonctions'. *Meddel. Lunds Univ. Math. Sem.*, **3**, 1–118. [6.2]

Fujiwara, M. & Kakeya, S. (1917). 'On some problems of maxima and minima for the curve of constant breadth and the in-revolvable curve of the equilateral triangle'. *Tôhoku Mathematical Journal*, **11**, 92–110. [7.1]

Garnett, J. (1970). 'Positive length but zero analytic capacity'. *Proceedings of the American Mathematical Society*, **24**, 696–9. [6.2]

Gillis, J. (1934a). 'On the projection of irregular linearly measurable plane sets of points'. *Proceedings of the Cambridge Philosophical Society*, **30**, 47–54. [6.4]

Gillis, J. (1934b). 'On linearly measurable plane sets of points of upper density $\frac{1}{2}$' *Fundamenta Mathematicae*, **22**, 57–69. [3.3]

Gillis, J. (1935). 'A theorem on irregular linearly measurable sets of points', *Journal of the London Mathematical Society*, **10**, 234–40. [2.2]

Gillis, J. (1936a). 'Note on the projection of irregular linearly measurable plane sets of points'. *Fundamenta Mathematicae*, **26**, 228–33. [6.4]

Gillis, J. (1936b). 'On linearly measurable sets of points'. *Comptes Rendus des Séances de la Societé des Sciences et des Lettres de Varsovie*, **28**, 49–70. [6.4]

Gołąb, S. (1929). 'Sur quelques points de la théorie de la longueur'. *Ann. Soc. Polon. Math.*, **7**, 227–41. [3.2]

Good, I.J. (1941). 'The fractional dimension theory of continued fractions'. *Proceedings of the Cambridge Philosophical Society*, **37**, 199–288. [8.5]

Grassberger, P. (1981). 'On the Hausdorff dimension of fractal attractors'. *Journal of Statistical Physics*, **26**, 173–9 [8.7]

Guckenheimer, J. & Holmes, P. (1983). *Nonlinear Oscillations, Dynamical Systems, and Bifurcations of Vector Fields*, New York: Springer.

de Guzmán, M. (1975). '*Differentiation of Integrals in* $\mathbb{R}^n$'. Berlin: *Springer Lecture Notes in Mathematics* **481**. [1.3, 7.2, 7.5]

de Guzmán, M. (1979). 'Besicovitch theory of linearly measurable sets and Fourier analysis'. In *Proceedings of Symposia in Pure Mathematics*, **35**, (1), *Harmonic Analysis in Euclidean Spaces*, pp. 61–7. Providence: American Mathematical Society. [7.2]

de Guzmán, M. (1981). 'Real variable methods in Fourier analysis'. Amsterdam: North Holland, *Mathematical Studies* **46**. [1.3, 3.1, 7.2, 7.5]

Haken, H. (1982). (Ed). *Evolution of Order and Chaos in Physics, Chemistry and Biology*. New York: *Springer Series in Synergetics*, **17**. [8.7]

Hausdorff, F. (1919). 'Dimension und äusseres Mass'. *Mathematische Annalen*, **79**, 157–79. [Int.]

Hawkes, J. (1974). 'Hausdorff measure, entropy and the independence of small sets'. *Proceedings of the London Mathematical Society* (3), **28**, 700–24. [Int.]

Hayman, W.K. & Kennedy, P.B. (1976). *Subharmonic Functions*, Volume 1. New York: Academic Press. [6.2]

Hénon, M. & Pomeau, Y. (1976). 'Two strange attractors with a simple structure'. In *Turbulence and Navier–Stokes Equations*, ed. R. Temam, pp. 29–68. *Lecture Notes in Mathematics* **565**, New York: Springer. [8.7]

Hida, T. (1980). *Brownian Motion*. New York: Springer-Verlag. [8.8]

Hille, E. (1973). *Analytic Function Theory*, Volume 2., 2nd ed. New York: Chelsea Publishing Company. [6.2]

Hirst, K.E. (1967). 'The Apollonian packing of circles'. *Journal of the London Mathematical Society*, **42**, 281–91. [8.4]

Hofstadter, D.R. (1981). 'Strange attractors: mathematical patterns delicately poised between order and chaos'. *Scientific American*, **245** (November), 16–29. [8.7]

Hurewicz, W. & Wallman, H. (1941). *Dimension Theory*. Princeton University Press. [Int.]

Hutchinson, J.E. (1981). 'Fractals and self-similarity'. *Indiana University Mathematics Journal*, **30**, 713–47. [8.3]

Ivanov, L.D. (1975). *Variatsii mnozhestv i funktsii* (Russian). Moscow: Nauka. [3.5]

Jarnik, V. (1931). 'Über die simultanen diophantischen Approximationen'. *Mathematische Zeitschrift*, **33**, 505–43. [8.5]

Johnson, R.A. & Rogers, C.A. (1982). 'Hausdorff measure and local measure'. *Journal of the London Mathematical Society* (2), **25**, 99–114. [Int.]

Jonker, L. & Rand, D. (1981*a*). 'Bifurcations in one dimension I. The non-wandering set'. *Inventiones Mathematicae*, **62**, 347–65. [8.7]

Jonker, L. & Rand, D. (1981*b*). 'Bifurcations in one dimension II. A versal model for bifurcations'. *Inventiones Mathematicae*, **63**, 1–15. [8.7]

Kahane, J.P. (1969). 'Trois notes sur les ensembles parfaits lineaires'. *L'enseignement Mathematique Revue Internationale* (2), **15**, 185–92. [7.2]

Kahane, J.P. (1976). 'Mesures et dimensions'. In *Turbulence and Navier–Stokes Equations*, ed. R. Temam, pp. 94–103. *Lecture Notes in Mathematics 565*, New York: Springer. [Int.]

Kakeya, S. (1917). 'Some problems on maxima and minima regarding ovals'. *Tôhoku Science Reports*, **6**, 71–88. [7.1]

Kametani, S. (1945). 'On Hausdorff's measures and generalized capacities with some of their applications to the theory of functions'. *Japanese Journal of Mathematics*, **19**, 217–57. [6.2]

Kaufman, R. (1968). 'On Hausdorff dimension of projections'. *Mathematika*, **15**, 153–5. [6.3]

Kaufman, R. (1969). 'An exceptional set for Hausdorff dimension'. *Mathematika*, **16**, 57–8. [8.5]

Kaufman, R. (1970). 'Probability, Hausdorff dimension and fractional distribution'. *Mathematika*, **17**, 63–7. [8.5]

Kaufman, R. (1981). 'On the theorem of Jarník and Besicovitch'. *Acta Arithmetica*, **39**, 265–7. [8.5]

Kaufman, R. & Mattila, P. (1975). 'Hausdorff dimension and exceptional sets of linear transformations'. *Annales Academiae Scientiarum Fennicae A*, **1**, 387–92. [6.3, 8.5]

Kingman, J.F.C. (1973). 'Subadditive ergodic theory'. *Annals of Probability*, **1**, 883–909. [8.4]

Kingman, J.F.C. (1976). 'Subadditive processes'. In *École d'Été de Probabilitiés de Saint-Flour V*, pp. 167–223. *Lecture notes in Mathematics* **539**, New York: Springer. [8.4]

Kingman, J.F.C. & Taylor, S.J. (1966). *Introduction to Measure and Probability*. Cambridge University Press. [1.1, 3.2, 6.2]

Kinney, J.R. (1968). 'A thin set of circles'. *American Mathematical Monthly*, **75**, 1077–81. [7.3]

Kinney, J.R. (1970). 'A thin set of lines'. *Israel Journal of Mathematics*, **8**, 97–102. [7.2]

Klee, V.L. (1957). 'Research problem no. 5', *Bulletin of the American Mathematical Society*, **63**, 419. [8.6]

Klee, V.L. (1959). 'Some characterizations of convex polyhedra'. *Acta Mathematica*, **102**, 79–107. [8.6]

Kline, S.A. (1945). 'On curves of fractional dimension'. *Journal of the London Mathematical Society*, **20**, 79–86. [8.2]

Lanford, O.E. (1982). 'A computer assisted proof of the Feigenbaum conjectures'. *Bulletin of the American Mathematical Society* (2), **6**, 427–34. [8.7]

Larman, D.G. (1966). 'On the exponent of convergence of a packing of spheres'. *Mathematika*, **13**, 57–9. [8.4]

Larman, D.G. (1967a). 'On Hausdorff measure in finite-dimensional compact metric spaces'. *Proceedings of the London Mathematical Society* (3), **17**, 193–206. [5.3]

Larman, D.G. (1967b). 'On the Besicovitch dimension of the residual set of arbitraily packed discs in the plane'. *Journal of the London Mathematical Society*, **42**, 292–302. [8.4]

Larman, D.G. (1967c). 'On the convex measure of product and cylinder sets'. *Journal of the London Mathematical Society*, **42**, 447–55. [5.3]

Larman, D.G. (1971a). 'A compact set of disjoint line segments in E^3 whose end set has positive measure'. *Mathematika*, **18**, 112–25. [7.2]

Larman, D.G. (1971b). 'On a conjecture of Klee and Martin for convex bodies'. *Proceedings of the London Mathematical Society* (3), **23**, 668–82. [8.6]

Lebesgue, H. (1904). *Leçons sur l'intégration et la récherche des fonctions primitives*. Paris: Gauthier-Villars. [Int.]

Lorenz, E.N. (1963). 'Deterministic nonperiodic flows'. *Journal of the Atmospheric Sciences*, **20**, 130–41. [8.7]

Love, E.R. & Young, L.C. (1937). 'Sur une classe de fonctionnelles linéaires'. *Fundamenta Mathematicae*, **28**, 243–57. [8.2]

Mandelbrot, B.B. (1975). *Les objects fractals: forme, hasard et dimension*. Paris: Flammarion. [Int.]

Mandelbrot, B.B. (1977). *Fractals: Form, Chance, and Dimension*. San Francisco: W.H. Freeman & Co. [Int., 8.2, 8.3]

Mandelbrot, B.B. (1982). *The Fractal Geometry of Nature*. San Francisco: W.H. Freeman & Co. [Int., 8.3, 8.7, 8.8, 8.9]

Mandelbrot, B.B. (1983). 'On discs and sigma discs, that osculate the limit sets of groups of inversions'. *Mathematical Intelligencer*, **5**, No. 2, 9–17. [8.4]

Marion, J. (1979). 'Le calcul de la mesure Hausdorff des sous-ensembles parfaits isotypiques de $\mathbb{R}^m$'. *Comptes Rendus Acad. Sci. Paris*, **289**, A65–A68. [8.3]

Marstrand, J.M. (1954a). 'Some fundamental geometrical properties of plane sets of fractional dimensions'. *Proceedings of the London Mathematical Society* (3), **4**, 257–302. [2.2, 4.1, 4.2, 4.3, 6.1, 6.3, 6.4, 6.5, 8.5]

Marstrand, J.M. (1954b). 'The dimension of Cartesian product sets'. *Proceedings of the Cambridge Philosophical Society*, **50**, 198–202. [5.1, 5.3]

Marstrand, J.M. (1955). 'Circular density of plane sets'. *Journal of the London Mathematical Society*, **30**, 238–46. [4.1, 4.3]

Marstrand, J.M. (1961). 'Hausdorff two-dimensional measure in 3-space', *Proceedings of the London Mathematical Society* (3), **11**, 91–108. [3.5]

Marstrand, J.M. (1964). 'The (ϕ, s)-regular subsets of n-space'. *Transactions of the American Mathematical Society*, **113**, 369–92. [3.5, 4.1, 4.4]

Marstrand, J.M. (1972). 'An application of topological transformation groups to the Kakeya problem'. *Bulletin of the London Mathematical Society*, **4**, 191–5. [7.4]

Marstrand, J.M. (1979a). 'Packing smooth curves in $\mathbb{R}^q$'. *Mathematika*, **26**, 1–12. [7.4]

Marstrand, J.M. (1979b). 'Packing planes in $\mathbb{R}^3$'. *Mathematika*, **26**, 180–3. [7.3]

Mattila, P. (1975a). 'Hausdorff m regular and rectifiable sets in n-space'. *Transactions of the American Mathematical Society*, **205**, 263–74. [3.5]

Mattila, P. (1975b). 'Hausdorff dimension, orthogonal projections and intersections with planes'. *Annales Academiae Scientiarum Fennicae A*, **1**, 227–44. [6.3, 6.5]

Mattila, P. (1981). 'Integralgeometric properties of capacities'. *Transactions of the American Mathematical Society*, **266**, 539–54. [6.5]

Mattila, P. (1982). 'On the structure of self-similar fractals'. *Annales Academiae Scientiarum Fennicae A*, **7**, 189–95. [6.5]

Mattila, P. (1984a). 'Hausdorff dimension and capacities of intersections of sets in n-space'. *Acta Mathematica*, **152**, 77–105. [6.5]

Mattila, P. (1984b). 'An integral inequality for capacities'. *Mathematica Scandinavica*, **53**, 256–64. [6.5]

Mattila, P. (1984c). 'A class of sets with positive length and zero analytic capacity. *Annales Academiae Scientiarum Fennicae*. To appear. [6.2].

May, R.M. (1976). 'Simple mathematical models with very complicated dynamics'. *Nature*, **261**, 459–67. [8.7]

McMinn, T.J. (1960). 'On the line segments of a convex surface in E_3'. *Pacific Journal of Mathematics*, **10**, 943–6. [8.6]

Mitjagin, B.S. & Nikisin, E.M. (1973). 'On the divergence of Fourier Series almost everywhere'. *Doklady Akademin Nauk SSSR*, **210**, 23–5. [7.5]

Moore, E.F. (1950). 'Density ratios and $(\phi, 1)$ rectifiability in n-space'. *Transactions of the American Mathematical Society*, **69**, 324–34. [3.5]

Moran, P.A.P. (1946). 'Additive functions of intervals and Hausdorff measure'. *Proceedings of the Cambridge Philosophical Society*, **42**, 15–23. [5.3, 8.3]

Moran, P.A.P. (1949). 'On plane sets of fractional dimensions'. *Proceedings of the London Mathematical Society* (2), **51**, 415–23. [5.3]

Morgan, G.W. (1935). 'The density directions of irregular linearly measurable plane sets'. *Proceedings of the London Mathematical Society* (2), **38**, 481–94. [2.2, 6.4]

Morse, A.P. & Randolph, J.F. (1944). 'The ϕ rectifiable subsets of the plane'. *Transactions of the American Mathematical Society*, **55**, 236–305. [3.3, 3.5]

Nikodym, O. (1927). 'Sur la mesure des ensembles plans dont tous les points sont rectalineairément accessibles'. *Fundamenta Mathematicae*, **10**, 116–68. [7.2]

Oberlin, D.M. & Stein, E.M. (1982). 'Mapping properties of the Radon transform'. *Indiana University Mathematics Journal*, **31**, 641–50. [7.3, 7.5]

Pál. J. (1921). 'Ein Minimumproblem für Ovale'. *Mathematische Annalen*, **83**, 311–9. [7.1, 7.4]

Patterson, S.J. (1976). 'The limit set of a Fuchsian group'. *Acta Mathematica*, **136**, 241–73. [8.4]

Pelling, M.J. (1977). 'Formulae for the arc-length of a curve in $\mathbb{R}^m$'. *American Mathematical Monthly*, **84**, 465–7. [3.2]

Perron, O. (1928). 'Über einen Satz von Besicovitch'. *Mathematische Zeitschrift*, **28**, 383–6. [7.1, 7.2]

Peyrière, J. (1977). 'Calculs de dimensions de Hausdorff'. *Duke Mathematical Journal*, **44**, 591–601. [1.5]

Pruitt, W.E. (1975). 'Some dimension results for processes with independent increments'. In *Stochastic Processes and Related Topics*, **1**, 133–65, ed. M.L. Puri, New York: Academic Press. [8.8]

Pruitt, W.E. & Taylor, S.J. (1969). 'Sample path properties of processes with stable components'. *Zeitschrift für Wahrscheinlichkeitstheorie*, **12**, 267–89. [8.8]

Putnam, C.R. (1974). 'The role of zero sets in the spectra of hypernormal operators'. *Proceedings of the American Mathematical Society*, **43**, 137–40. [7.5]

Rademacher, H. (1962). 'On a theorem of Besicovitch'. In *Studies in Mathematical Analysis–Essays in Honour of G. Pólya*, pp. 294–6. Stanford University Press. [7.2]

Randolph, J.F. (1936). 'On generalizations of length and area'. *Bulletin of the American Mathematical Society*, **42**, 268–74. [5.3]

Randolph, J.F. (1941). 'Some properties of sets of the Cantor type'. *Journal of the London Mathematical Society*, **16**, 38–42. [1.5]

Ravetz, J. (1954). 'The Denjoy theorem and sets of fractional dimension'. *Journal of the London Mathematical Society*, **29**, 88–96. [1.5]

Ray, D. (1963). 'Sojourn times and the exact Hausdorff measure of the sample path for planar Brownian motion'. *Transactions of the American Mathematical Society*, **106**, 436–44. [8.8]

Reifenberg, E.R. (1960). 'Solution of the Plateau problem for m-dimensional surfaces of varying topological type'. *Acta Mathematica, Stockholm*, **104**, 1–92. [3.5]

Rogers, C.A. (1970). *Hausdorff Measure*. Cambridge University Press. [Int., 1.1, 1.2, 3.2, 5.1, 5.2, 8.5]

Rogers, C.A. et al. (1980). *Analytic Sets*. New York: Academic Press. [1.1]

Rudin, W. (1970). *Real and Complex Analysis*, New York: McGraw Hill. [6.2]

Rudin, W. (1973). *Functional Analysis*. New York: McGraw Hill. [6.2]

Ruelle, D. (1983). 'Bowen's formula for the Hausdorff dimension of self-similar sets'. In *scaling and self-similarity in physics-renormalization in statistical mechanics and dynamics*, Progress in physics, **7** Birkhauser [8.7]

Santaló, L.A. (1976). *Integral Geometry and Geometric Probability*. Reading, Massachusetts: Addison-Wesley. [6.4, 6.5]

Schmidt, W.M. (1980). *Diophantine Approximation*. Berlin: Springer, *Lecture Notes in Mathematics*, **785**. [8.5]

Schoenberg, I.J. (1962a). 'On certain minima related to the Besicovitch Kakeya problem'. *Mathematica (Cluj)*, **4** (27), 145–8. [7.2]

Schoenberg, I.J. (1962b). 'On the Besicovitch–Perron solution of the Kakeya problem'. In *Studies in Mathematical Analysis–Essays in Honour of G. Pólya.*, pp. 359–63. Stanford University Press. [7.2]

Simó, C. (1979). 'On the Hénon-Pomeau attractor'. *Journal of Statistical Physics*, **21**. 465–94. [8.7]

Smale, S. (1967). 'Differentiable dynamical systems'. *Bulletin of the American Mathematical Society*, **73**, 747–817. [8.7]

Soddy, F. (1936). 'The kiss precise'. *Nature*, **137**, 1021. [8.4]

Sparrow, C. (1982). *The Lorenz Equations: Bifurcations, Chaos, and Strange Attractors*. New York: Springer-Verlag. [8.7]

Stein, E.M. (1976). 'Maximal functions; spherical means'. *Proceedings of the National Academy of Sciences of the USA*, **73**, 2174–5. [7.5]

Stein, E.M. & Wainger, S. (1978). 'Problems in harmonic analysis related to curvature'. *Bulletin of the American Mathematical Society*, **84**, 1239–95. [7.5]

Steinhaus, H. (1920). 'Sur les distances des points des ensembles de mesure positive'. *Fundementa Mathematicae*, **1**, 93–104. [7.4]

Stepney, S. (1984). 'Snowflakes and other monsters'. *Acorn User*, March 1984. [8.3]

Sullivan, D. (1979). 'The density at infinity of a discrete group of hyperbolic motions'. *Institut des Hautes Études Scientifiques. Publications Mathematics*, **50**, 171–202 [8.4]

Talagrand, M. (1980). 'Sur la measure de la projection d'un compact et certaines familles de cercles'. *Bulletin Sciences et Mathematiques*, **104**, 225–31. [7.3]

Taylor, S.J. (1952). 'On Cartesian product sets'. *Journal of the London Mathematical Society*, **27**, 295–304. [5.3]

Taylor, S.J. (1953). 'The Hausdorff α-dimensional measure of Brownian paths in *n*-space'. *Proceedings of the Cambridge Philosophical Society*, **48**, 31–9. [8.8]

Taylor, S.J. (1961). 'On the connection between Hausdorff measures and generalized capacities'. *Proceedings of the Cambridge Philosophical Society*, **57**, 524–31. [6.2]

Taylor, S.J. (1964). 'The exact Hausdorff measure of the sample path for planar Brownian motion'. *Proceedings of the Cambridge Philosophical Society*, **60**, 253–8. [8.8]

Taylor, S.J. (1973). 'Sample path properties of processes with stationary independent increments'. In *Stochastic Analysis*, eds. Kendall, D.G. & Harding, E.G., J. Wiley & Sons. [8.8].

Taylor, S.J. (1975). 'Abram Samoilovitch Besicovitch (obituary)'. *Bulletin of the London Mathematical Society*, **7**, 191–210. [Int., 7.4]

Temam, R. (1976) (ed.). *Turbulence and Navier–Stokes Equations.* New York: *Springer Lecture Notes in Mathematics*, **565**, [8.7]

Temam, R. (1983). *Navier–Stokes Equations and Non-linear Functional Analysis.* Philadelphia: Society for Industrial and Applied Mathematics. [8.7]

Tricot, C. (1981). 'Douze definitions de la densité logarithmique'. *Comptes Rendus de séances de l'académie des sciences de Paris* (1), **293**, 549–52. [Int.]

Tricot, C. (1982). 'Two definitions of fractional dimension'. *Mathematical Proceedings of the Cambridge Philosophical Society*, **91**, 57–74. [Int.]

Ugaheri, T. (1942). 'On the Newtonian capacity and the linear measure'. *Proceedings of the Imperial Academy of Japan*, **18**, 602–5. [6.2]

Ville, A. (1936). 'Ein satz über quadratische lange, Ergebrisse eins math'. *Kolloquiums* (*Wien*), **7**, 22–3. [8.2]

Vitushkin, A.G. (1966). 'A proof of the upper semicontinuity of a set variation'. (Russian). *Doklady Akademin Nauk SSSR*, **166**, 1022–5. Translation: *Soviet Math. Dokl*, **7** (1966), 206–9. [3.5]

Wainger, S. (1979). 'Applications of Fourier transforms to averages over lower dimensional sets'. In *Proceedings of Symposia in Pure Mathematics*, **35**, ed. S. Wainger & G. Weiss, pp. 85–94. Providence: American Mathematical Society. [Ex. 7.5]

Wainger, S. & Weiss, G. (1979) (eds.). 'Harmonic analysis in Euclidean spaces'. *Proceedings of Symposia in Pure Mathematics*, **35**, Providence: American Mathematical Society. [7.5]

Walker, R.J. (1952). 'Addendum to Mr Greenwood's paper'. *Pi Mu Epsilon Journal*, **275**. [7.4]

Wallin, H. (1969). 'Hausdorff measures and generalized differentiation'. *Mathematische Annalen*, **183**, 275–86. [2.2, 6.2]

Ward, D.J. (1967). 'The measure of cylinder sets'. *Journal of the London Mathematical Society*, **42**, 401–8. [5.3]

Ward, D.J. (1970a). 'A set of plane measure zero containing all finite polygonal arcs'. *Canadian Journal of Mathematics*, **22**, 815–21. [7.4]

Ward, D.J. (1970b). 'Some dimensional properties of generalized difference sets'. *Mathematika*, **17**, 185–8. [7.4]

Watson, G.N. (1966). *A Treatise on the Theory of Bessel Functions.* 3rd. ed. Cambridge University Press. [6.3]

Ważewski, T. (1927). 'Rectifiable continua in connection with absolutely continuous functions and mappings'. (Polish). *Ann. Soc. Polon. Math.,* **3**, 9–49. [3.2]

Wegmann, H. (1969*a*). 'Die Hausdorff-Dimension von kartesischen Produktmengen in metrischen Ráumen'. *Journal fur die Reine und Angewandte Mathematik,* **234**, 163–71. [5.3]

Wegmann, H. (1969*b*). 'Die Hausdorff-Dimension von Mengen reeler Zahlenfolgen'. *Journal fur die Reine und Angewandte Mathematik,* **235**, 20–8. [5.3]

Wegmann, H. (1971*a*). 'Die Hausdorff-Dimension von kartesischen Produkten metrischer Räume'. *Journal fur die Reine und Angewandte Mathematik,* **246**, 46–75. [5.3]

Wegmann, H. (1971*b*). 'Das Hausdorff-Mass von Cantormengen'. *Mathematische Annalen,* **193**, 7–20. [1.5]

Wetzel, J.E. (1973). 'Sectorial covers for curves of constant length'. *Canadian Mathematical Bulletin,* **16**, 367–75. [7.4]

Whitley, D. (1983). 'Discrete dynamical systems in dimensions one and two'. *Bulletin of the London Mathematical Society,* **15**, 177–217. [8.7]

Wiener, N. (1923). 'Differential space'. *Journal of Mathematical Physics,* **2**, 131–74. [8.8]

Wilker, J.B. (1971). 'The non-Euclidean Kakeya problem'. In *Proceedings of the 25th Summer Meeting of the Canadian Mathematical Congress,* pp. 603–7. [7.20]

Index Entries in bold type refer to definitions